John Wallace Turner

1987.

# LANDSCAPES OF STONE

*J.B.Whittow*

# LANDSCAPES OF STONE

WHITTET BOOKS

First published 1986
© 1986 by J.B. Whittow
Whittet Books Ltd, 18 Anley Road, London W14 0BY
Design by Paul Minns
All rights reserved

British Library Cataloguing in Publication Data

Whittow, John
    Landscapes of stone.
    1. Geology—Great Britain
    I. Title
    554.1        QE261

    ISBN 0–905483–50–2

Typeset by Inforum Ltd, Portsmouth
Printed and bound by the Bath Press, Bath

*To Diane, for her constant encouragement*

# Contents

# Acknowledgments

The author is greatly indebted to the following who have assisted with the preparation of this book: Sarah Prentice who typed the MS and contributed the line-drawings; Christopher Howitt who drew the maps and diagrams; Erika Meller who assisted with the photographic reproduction; Brian Knapp, Bob Parry, Jenny Jones, Harry Walkland, Suzanne Watson, Roger Farrington and Spink and Son Ltd of London who loaned photographs. Grateful thanks are also given to Moyra Stewart, Richard Saull, Chris Holland and Monique Giles who helped with the index.

The author and publishers gratefully acknowledge the permission of Andrew Goudie and Rita Gardner upon whose diagrams (in *Discovering Landscape in England and Wales*, Allen and Unwin) the diagrams 8a and 8b on p. 79 were based.

All illustrations not otherwise credited are taken by the author. The author and publishers gratefully acknowledge permission to reproduce the illustrations that appear on the following pages by these people: Aberdeen University Library, G.W. Wilson Collection, p. 139; British Tourist Authority, pp. 45, 46, 77, 127, 131, 136, 167, 168; Caroline Forbes, pp. 80 (2), 82, 86, 87; Fox Photos, p. 60; Institute of Agricultural History and Museum of English Rural Life, University of Reading, pp. 38, 60, 92, 94; Brian Knapp, pp. 31, 58, 59, 130, 132, 155, 156; Richard Sale, pp. 58, 94, 118, 130, 153; Scottish Tourist Board, pp. 66, 68, 135, 178; Edwin Smith, pp. 41, 42, 63, 64, 74(2), 75, 79, 82, 83, 84, 91, 99, 102, 104, 110, 111, 113, 115, 119, 126, 128, 138, 140(2); Spink and Son, p. 85; Charlie Waite, pp. 52, 62.

# INTRODUCTION

At first glance stone appears to be inert, immovable and intractable. Almost everywhere solid rock forms the skeleton of our landscape, the framework upon which the delicate skin of the cultural scene has been moulded. Yet this apparent immobility is an illusion – landscape is dynamic. It changes from place to place, it evolves over time, it is transformed through the seasons and with different moods of weather. Although nature has fashioned the rocks which build our 'everlasting hills' its task is never done. In viewing our scenery one of the earliest geologists (James Hutton in 1788) saw no vestige of a beginning, no prospect of an end. The natural processes of frost, wind, rain and running water are even now etching out the intricate details of our mountain slopes and river valleys and will continue to do so. Chunks of stone are constantly being prised by frost from the upland cliffs, whence they are scoured by wind and pitted by rain before being rolled and bounced by streams which bludgeon them into pebbles. The slow journey from uplands to lowlands is a stuttering progression, with years of immobility being suddenly punctuated by short periods of frenzied activity. These are the times of storm and tempest, when landslides scar the hillsides and streams rampage in spate. The peaceful glens and dales, with their memories of sunlit picnics alongside murmuring or chattering brooks, are now converted into clamorous conduits through which roaring torrents funnel countless tons of rocks and soil down to the lowland estuaries and thence to the sea. Here is the final repository where virtually all our layered rocks are first created, deep down below the seas and oceans.

It is important to realize that some rocks are formed in other ways, as will be shown in Chapter 1, but it is the bedded sedimentary rocks, formed from thick layers of sediment brought down from mountain to coast on nature's conveyor belt, that have played the greatest role in our scenic heritage. Although granites build the Dartmoor tors and ancient lavas create the highest summits of England, Scotland and Wales, these are exceptional landforms. More typical are the Pennines' gritstone moors, the sandstone ridges of Central Scotland, South Wales and Midland England, the limestone scarps of Shropshire and Gloucester-shire, the chalk downs of Wiltshire or Sussex. These are our everyday land-scapes, naturally sculptured into neighbouring hill and familiar valley. Only one other operation has been as dynamic or as important as these natural processes in the fashioning of the British landscape and, not surprisingly, this involves the work of man.

From very earliest times mankind has realized the value of stone, whether as a basic implement such as an axe, as a weapon such as a sling-shot, or as a means of primitive shelter. When our forebears finally left their caves to construct the earliest dwellings, one of the first uses entailed no more than selecting suitable pebbles from the stony hillsides, river beds or shingle beaches. Where rocks had already split naturally into slabs (flags) the local tribes were able to use them in simple building construction, but elsewhere in the rocky uplands the prehistoric dwellings were little more than heaps of loose boulders rudely fashioned into circles and roofed by heather thatch. In the wilderness areas of Scotland vernacular architecture had not progressed very far from these primitive beginnings when the twentieth century dawned. But elsewhere, the introduction of metal technologies saw prehistoric man make significant changes in Britain's cultural scene. No longer restricted by relatively ineffective stone tools, the population began to make substantial inroads into the hitherto impenetrable forests of lowland Britain. Not only were they able to expand their farming practices on to the more fertile soils but they were also able to utilize the forest timber and the clay from newly cleared vales to construct more substantial dwellings. The wattle and daub eventually gave way to oak frame and brick in the formerly forested lowlands, although the traditional methods of building with stone persisted for centuries in the bare mountains and moorlands. This basic dichotomy of vernacular architecture survived in Britain from Roman times until the eighteenth century, a division that is reflected partly, but not wholly, in the contrasting landscape of highland and lowland Britain. Our scenic heritage can be likened to a complex mosaic pavement laid down as a thin veneer on a base of stonework. In some places the tiny pieces are missing, allowing the stone base to show through, as in our mountainlands and some sea coasts. In most places the muted colours and textures of the mosaic's rock slivers have been so carefully matched that they create a harmonious composition; these are the unspoiled rural landscapes where man and nature work happily as equal partners in an antique land. Elsewhere, however, the ancient mosaic has been destroyed, only to be replaced by garish modern materials that fail to mix with the original colours and textures; such are the modern urban sprawls of concrete, steel and brick. The Industrial Revolution, together with the high technology of modern Britain, for all their achievements, have set about dismembering the delicate mosaic of landscape patterns that have taken millennia to evolve.

This book tries to capture something of that rapidly disappearing mosaic before it is lost, just as the tessellated pavements of Roman Britain were buried by the clutter of succeeding civilizations. There are many existing books which describe and explain the British scene; the hedgerows of the lowlands and the stone walls of the uplands; the half-timbered hamlets of Midland England and the gnarled stone cottages of Scotland or the Lakes; the pantiles and flint of East Anglia and the slated roofs of Wales. This book is different. It looks only at *Landscapes of Stone*; stone is the most durable of materials, whether it be part of a mountain top, escarpment, fieldwall, barn, mansion or cottage. Stone sits easily in the landscape because it belongs most naturally to it. It makes no matter if the stone is a cold, grey granite of a Cornish fishing village or a warm, creamy limestone of a Cotswold farmstead, because both have such a close affinity with their respective environments that they seem to have grown out of the ground itself, which in a sense they have. For all the mellowness of some ancient bricks and tiles they are essentially materials of the modern city. Stone, on the other hand, because of its bulk, was used mainly where it was found, its different colours and textures bringing subtle variations of character to the rural scene. It is true that some of the finest freestones were carried long distances to build the great churches and cathedrals of the Middle Ages or to enhance the stately homes of Georgian Britain, but it was not until the advent of the canals and railways that building stone could be quarried and transported to help build cities all over the realm. At once the closely knit geological and environmental affinities were broken. After Victorian times the harmonious links between buildings and landscape were never quite the same – some of that precious sense of place had been lost for ever.

PART I

# *IN THE BEGINNING*

# *THE GEOLOGICAL FOUNDATION*

The richly varied landscapes of Britain may have been described as having stimulated our finest painting, music and literature but how many of the artists, composers and novelists had a real understanding of the intricately detailed structures that underlie the scenery which they have so delicately and feelingly portrayed? There seems little doubt that many of them must have had their natural curiosity aroused, for how can anyone fail to ask questions when confronted with the contorted rocks of Cornish sea cliffs, when faced with the grey limestone walls of Cheddar Gorge and Dovedale, or when first setting eyes on the mist-shrouded peaks of Snowdonia and the Lake District? To fully understand these differences in landscape it is necessary to know, for example, how rocks differ from place to place, why granite will create a distinctive type of upland, how chalk has been fashioned into rolling downlands and why certain rocks make better building stones than others. Before looking in more detail at the rock-forming processes, however, it would be helpful to say something about the geology map.

Even a cursory glance at a professionally published map of Britain will illustrate the complexity of our geological foundation. Few countries exhibit such a diversity of rock types in so small an area. It is impossible to travel far without crossing a geological boundary and it is this irregular juxtaposition of contrasting rocks that gives to the British scene its remarkable variety. The geology map can be compared with a multicoloured patchwork quilt rather than a finely woven tapestry, for at first glance, its constituents appear to have no symmetry, only a rough and ready mixture of apparently unrelated patterns. Yet there are relationships within the jumble and a framework can be discerned if given a little careful scrutiny. It will be found that in general the older, harder rocks occur in the west and north, while the younger, less resistant rocks are confined mainly to the south and east. This broad division is what distinguishes Upland from Lowland Britain, each part having its distinctive set of landscape patterns. The geology map also illustrates that all rock types are classified into three broad groups according to their mode of formation: sedimentary, igneous and metamorphic (see Figure 1).

Figure 1

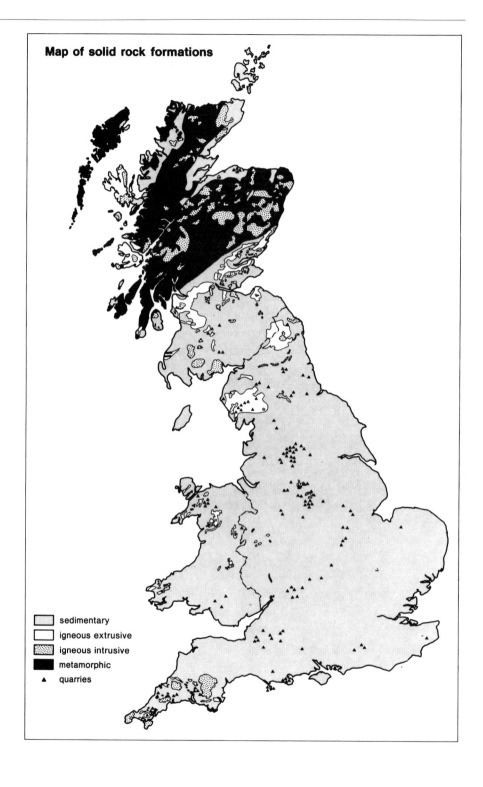

**Map of solid rock formations**

sedimentary
igneous extrusive
igneous intrusive
metamorphic
▲ quarries

# The sedimentary rocks

These, as their name suggests, are formed by the slow compression and solidification of sediments that have accumulated on the land surface or beneath the oceans. Geologists have separated them into two major divisions: first, those formed from materials which have been carried varying distances before being deposited; and second, those sediments which, instead of being transported, have accumulated *in situ* as dead plant and animal remains before they themselves are converted into solid rocks by chemical processes and compaction.

The first group, the so-called 'transported sediments', begin their lengthy journey when existing rocks are slowly broken up by frost, rain and wind, to produce a steady supply of loose stone fragments ranging in size from boulders and pebbles down to tiny, individual mineral grains. Slowly but surely these unconsolidated materials slither and slide down the hillslopes before finding their way into the network of streams which characterize all but desert environments. Running water speeds up their journey, with the stream currents rolling the larger pebbles along the river bed, bouncing the gravel and carrying the fine-grained silts and muds along in suspension. Not surprisingly, the bumping and battering suffered in the river channel reduces the size of the moving pebbles and smooths out their angularities. Wherever the river enters a lake or ocean its current slackens and most of its burden is dumped. The bigger stones are abandoned first, nearest to the shorelines, where the pebbles will finally be compacted into conglomerates; the coarse quartz grains are carried farther out, ultimately to be converted into sandstones, while the smallest and lightest particles are carried into deeper water where they sink to the sea floor before being transformed into mudstones, clays and shales.

Because natural processes transport material incessantly from the uplands to the lowlands and thence to the oceans one can understand why continents are being slowly worn down and nearshore ocean basins are being filled up. This is how the majority of sedimentary rocks are created, layer upon layer, each bed being separated from its neighbour by a flat surface known as a bedding plane. The beds of rocks, termed 'strata', are also interrupted by vertical cracks known as 'joints' which form at right angles to the bedding planes, thereby creating a type of lattice work within the solid rock. In sedimentary rocks the joints result from shrinkage when the sediment dries out, but these lines of weakness will eventually prove to be chinks in the rock's seemingly impregnable armour. It will be shown how joints and similar fractures can also be formed in igneous and metamorphic rocks (for different reasons) and how first nature and then quarrymen have made use of weaknesses to split, wedge and prise the stone blocks away from the 'living rock'. Nature's tools of frost and rain may take longer than the artifacts of man to chisel into a rock face but human efforts are puny when compared with the age-old processes that have sculpted the world's landscapes. Britain's landscapes of stone were being fashioned countless millions of years before man even appeared on the scene.

Sandstone is one of the commonest sedimentary rocks but, apart from the fact that all varieties are composed of colourless or whitish quartz grains, sandstones are found to display a whole range of colours and textures. White and pale buff-coloured stone occurs because it is virtually free of the iron mineral known as haematite. Most commonly, however, iron oxide has stained the majority of sandstones into every shade of pink, red, orange and brown. Apart from their colour differences, sandstones can also be differentiated according to the size and shape of their sugar-like grains. For instance, some sandstones incorporate coarse pebbles which make them difficult to work and restrict their building use largely to rubble walling of modest cottages and barns or to drystone fieldwall construction. Conversely, the finest building stones, termed '*freestones*' because they can be worked more freely, exhibit smaller, more uniformly sized grains which give a smoothly textured finish to cathedral and mansion alike. In general, windblown desert sands produce the best sorted and most even-textured sand-stones which the quarryman fashions into the finest smooth-faced masonry known as 'ashlar.' If one cares to use a magnifying glass it will be seen that the individual quartz grains are rounded instead of being angular. By contrast, where the quartz grains have not been transported far they retain their angular-ity, and these are what form a rougher textured sandstone known as 'gritstone'. It is because of these tough, angular quartz grains that the pale Millstone Grit of the Pennines has such magnificent grinding properties.

The finer-grained sediments, such as mudstones and shales, are made up of various proportions of clay, lime and decayed vegetable material termed 'carbo-naceous matter.' A high proportion of lime produces a light-coloured stone while a preponderance of decayed carbonaceous matter gives much darker tones to the rock, as in the case of coal. If the rock has no carbonaceous material but contains equal proportions of clay and lime, it is termed a 'marlstone' (or 'cementstone') which has been used in its crushed state for both agricultural and industrial purposes. Despite millions of years of compaction on the ocean floor, clays and mudstones are soft enough to be worn down quickly once they have been thrown up clear of the ocean. It is hardly surprising, therefore, that wherever these poorly resistant rocks occur, weathering has fashioned them into plains and vales. Only where they are toughened by interbedded sandstones, as in central

Wales and southern Scotland, have mudstones and shales made anything more than a minor contribution to the British uplands. Furthermore, because they yield quickly to weathering, clays and shales have no value as a building stone although their comparative softness allows them to be easily excavated for brick-making. The paucity of good building stone in south-east England reflects the high proportion of clays and soft limestones that form the region. Were it not for the widespread occurrence of flint, the traditional architecture of the south-east would be limited to timber and brick, except where a few sandstones occur in the Weald of Kent.

Of the non-transported sedimentary rocks, only the limestones and chalk have had any significant influence on both landforms and on building materials. These pale rocks make their mark mainly in England, for throughout Wales and Scotland, calcareous rocks are not so widespread and, with a few notable exceptions, play only a minor role in the true Highland landscapes. Although coal seams have contributed uniquely and overwhelmingly to some of Britain's most distinctive industrial landscapes, coal as a rock makes virtually no contribution to natural scenery.

Limestones differ fundamentally from sandstones and shales because, instead of being made up of rock particles, they are composed almost entirely of the shells and skeletons of sea creatures. When such organisms die their remains settle on the sea floor, but it must be remembered that some creatures actually build a 'living' type of rock in the form of a coral-reef limestone in certain tropical seas. Such marine creatures live most happily in warm oceans where food supplies are abundant and where they are far-removed from the sandy contamination of rivers. Exposures of shelly and coral limestones can be seen in such famous beauty spots as Wenlock Edge, the Mendips, Malham, the Gower and the Great Orme, Llandudno. A close inspection of the naked stone will reveal the presence of fossil shells; some shelly limestones are so attractive that they have been quarried for ornamental purposes, especially for use in church interiors. In marked contrast to these hard, greyish limestones is the soft, white limestone known as chalk, for it is so even-textured and structureless that its minute fossils can be distinguished only by means of a magnifying glass. Chalk is really a calcareous mud of considerable purity and dazzling whiteness, testifying to its lack of contamination and mineral staining. Characteristically, it forms gently rolling downlands except where it is truncated by the sea to produce the famous white bastions overlooking the Straits of Dover.

Visitors to the stone-built Cotswold villages may have noticed that this mellow limestone has a slightly grainy feel, not because it is made up of broken shells and corals but because it was formed from millions of tightly cemented limy pellets known as 'ooliths'. Because these resemble the roe of a fish (oolith = stone egg) the rock is referred to as 'oolitic limestone.' Like many of the fine-grained sandstones, these oolites (as they are commonly known) form excellent freestones, much sought after for building purposes because of their colour, even texture and ease of working. The durability and quality of their ashlar stone is exemplified not only by the Gothic cathedrals of southern England, such as Wells and Gloucester, but also by the way in which their natural outcrops have created the steadfast eminences of Cotswold scarp and Dorset sea cliff. As a rule limestone is white, as in the case of Portland's famous stone, but different impurities cause deeper shades: the honey-coloured Bath stone, the gingery-brown Northampton stone, the dove-greys of the Pennines, the blue-greys of Bristol, or the black 'marble' of Purbeck.

A different type of impurity in limestone has created the distinctive stone known as flint. This glistening black material, which splinters like glass into razor-sharp sherds, is a mass of pure quartz-like minerals known as 'silica'. In the clear seas where chalk was formed the few sand grains that were present changed into a silica-rich fluid before becoming amalgamated into a nodule, often around the skeleton of a sea urchin. Such a transformation can be likened to the melting of sugar crystals before their subsequent freezing into an ice lolly, although in the case of flint the silica crystals become fused rather than frozen. In its undisturbed state flint occurs either as irregular lumps or as a thin sheet interbedded in the solid chalk. Where the rocks have been eroded, pieces of broken flint have been rounded by water into pebbles, just like those on Brighton beach. Mixtures of angular and rounded flints have often become incorporated into younger deposits from which they have been dug for use as a building material throughout East Anglia and south-east England. Capping the Chilterns and the North Downs, for example, is a sticky, flinty clay termed clay-with-flints, the distribution of which is shown in Figure 5. It will be shown in Chapter 5 how the flinty Hertfordshire Pudding Stone received its name and why the grey rocks known as 'sarsens' litter parts of the Wessex downlands.

## The igneous rocks

In contrast to the way in which most sedimentary rocks have been formed beneath the oceans, the igneous rocks were spawned in the bowels of the earth. Igneous means 'fire-formed' and all rocks of this type have been derived from the molten material which seethes beneath the crust and occasionally bursts out at the surface from a volcanic vent or crack. Today Britain has no active volcanoes,

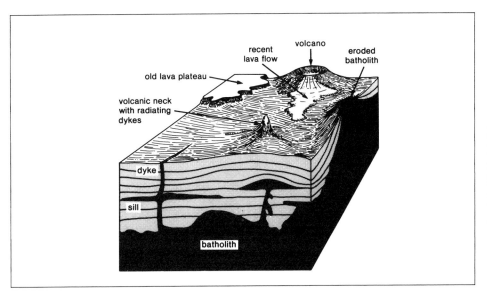

recent
lava flow
volcano
eroded
batholith
old lava plateau
volcanic neck
with radiating
dykes
dyke
sill
batholith

Figure 2
Igneous rock
formation,
showing the
different
landforms
produced by
intrusive
(black) rocks
and extrusive
(white) rocks.

although periodically volcanic explosions have rent the peaceful British scene: some 50 million years ago volcanoes towered where the isles of Mull and Arran now stand; Edinburgh itself is sited on the remnants of a huge volcano that once spewed forth ash and lava; the highest peaks of Scotland, England and Wales are all carved from volcanic rocks generated by even older volcanoes. Not all igneous rocks are of volcanic origin, however, for some of the molten material never broke through the crust but cooled more slowly at depth to form such rocks as the granites. Because they remained in the underworld such deep-seated rocks have been termed 'plutonic' or 'intrusive' rocks. They contrast with the differently formed volcanic rocks which poured out at the surface, hence their title of 'extrusive' rocks (see Figure 2).

Volcanic rocks were initially expelled either as molten lava or as billowing clouds of ash, pumice, cinder or simply as rock fragments shattered by the explosion. It is difficult to think of parts of Britain being overwhelmed by extensive lava flows like those of modern Iceland, but the plateaux of northern Skye, Mull and those that form the flat-topped hills known as 'tablelands' around Glasgow are built from layer upon layer of ancient lava, most of which is dark-coloured because it contains a high proportion of iron minerals. Basalt is the most common of the Scottish lavas, a rock characterized not only by its tiny crystals due to rapid cooling, but also by its hexagonal columnar structure which weathers into a distinctive 'organ-pipe' rock exposure. These well jointed basaltic columns are best seen at the legendary Fingal's Cave on the lonely

Hebridean island of Staffa. The strongly developed jointing also allows basalt to be easily quarried as a durable building stone, but because of its sombre colouring it is rare to find a Hebridean cottage without its coat of whitewash. Don't expect to see organ-pipe structures on every lava flow, more commonly the horizontal lavas have been eroded into featureless tablelands or simply stepped moorland landscapes.

Not all lavas have a high proportion of dark iron minerals since those that have a preponderance of silica are lighter in colour. These yellowish and pinkish lavas were more common in the older volcanoes which erupted in Snowdonia and the Lake District some 500 million years ago. But, unlike the Scottish basalts which flowed out at the surface, these ancient silica-rich lavas were extruded intermittently from submarine volcanoes. Thus, their flows became interbedded with ordinary sea floor sediments, as did their accompanying layers of powdery volcanic ash which were ultimately transformed into layers of hard rock. Lavas and ashes became sandwiched between beds of sedimentary rocks like a gigantic layer cake. This unappetizing confection was soon to form the basis of a new British landmass for no sooner had earth movements bent the layered rocks into enormous folds and uplifted them far above King Neptune's realm, than wind and rain commenced their demolition work on the puckered crust. It will be seen in Chapter 4 how the eroded layers have helped to give a rugged splendour to the Lake District. The toughness of these igneous rocks was soon discovered by early Britons who set up their stone-axe factories in North Wales and Cumbria.

Such convulsions and associated volcanic activity occurred periodically throughout geological time and at the close of each event the molten material remaining in the subterranean lava chambers slowly cooled and solidified. This was the time when such coarsely crystalline rocks as granite and gabbro were created. The lengthy cooling period allowed time for their characteristically large crystals to form deep in the earth.

Granites are almost invariably light-coloured because they are composed essentially of colourless quartz crystals together with two other minerals known as mica and feldspar. In fact the granite's overall colour is determined by the type of feldspar present: it may be red, pink or grey, thereby explaining why the silvery-grey stones of Aberdeen, the Granite City, contrast with the pink granite of Shap in Cumbria. Although the granite mass may extend to several hundred square miles at depth, its surface exposure may be considerably smaller. Such domed mountains as the Cairngorms and the Cheviots or the wild moorland of Dartmoor are mere fractions of the underlying chambers of granite, for their covering of sedimentary rocks has not been completely removed by erosion. Every bit of granite that you see in South-West England, for example, from

Dartmoor to the Isles of Scilly, belongs to the same underground mass, which stretches for some 150 miles.

Not all plutonic rocks are granitic, because of the different chemistry of the various molten materials. Although it has been seen how basalt cools rapidly as a surface lava flow to form a finely grained dark rock, at depth it will cool more slowly to create a coarsely crystalline, dark plutonic rock known as 'gabbro'. The Black Cuillins on the Scottish Isle of Skye are made of gabbro – their jagged outlines contrasting in both colour and form with the neighbouring granitic Red Hills. Dolerite is another common igneous rock, whose crystals are smaller than those of the gabbros but larger than the crystals of the extrusive lavas, thereby suggesting that dolerite must have cooled and solidified somewhere near the surface but without actually breaking through. When molten rock from the fiery subterranean chambers attempts to make its way to the surface other than through the vent of a volcano it will force itself along lines of weakness in the crust (see Figure 2). In some cases it squeezes along the bedding planes of the layered sedimentary rocks before cooling into a solid sheet of igneous material termed a 'sill'. Once uncovered by weathering this tough layer of dark rock will survive, even though the softer sedimentary rocks are worn away, often to form a cliff or step on the hillside over which waterfalls may plunge. The Great Whin Sill of northern England is one of the best known examples. Alternatively, the molten lava will rise to the surface along a vertical crack which may extend laterally for scores of miles. Once this intrusive band of igneous rock becomes solidified it will form a dyke. Like the sill, the dyke's rock is so resistant that it sometimes protrudes as a narrow wall of dark stone, usually less than 10 feet in width, running across the landscape. It is commonly seen on the sea shores and mountainsides of western Scotland.

## The metamorphic rocks

The third major group of rocks has been termed 'metamorphic' because it refers to rock that has been severely altered from its previous type (both sedimentary and igneous) by the effects of intense pressure and/or heat.

The simplest type of metamorphic change occurs when the lowest layers of sedimentary rocks become progressively buried by thick beds of new sediments. Partly because of the increasing pressure and partly because temperatures within the earth's crust rise with depth, new minerals will form to replace the older ones (a process known as re-crystallization), thereby creating a whole new rock type (see Figure 3). In a sandstone, for instance, the quartz grains are squashed so

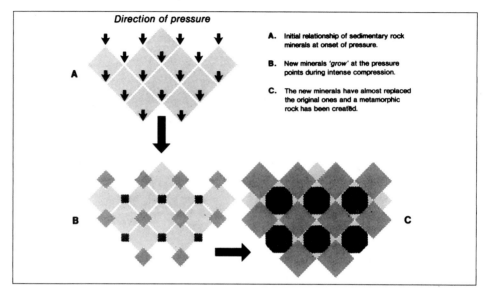

**Direction of pressure**

A.  Initial relationship of sedimentary rock minerals at onset of pressure.

B.  New minerals 'grow' at the pressure points during intense compression.

C.  The new minerals have almost replaced the original ones and a metamorphic rock has been created.

Figure 3
The process of recrystallization which turns a sedimentary rock into a metamorphic rock.

closely together that they become virtually 'fused' into a metamorphic quartzite, an extremely hard, glistening white stone which helps build some of Scotland's shapeliest peaks. Limestones too are compacted to such an extent that their shelly structure is obliterated and a marble is formed. The purest marbles are white but impurities of iron or copper will produce pink or green shades, respectively. Some of the so-called 'marbles' of Dorset, Sussex and the Cotswolds are not metamorphic rocks but pretty, patterned limestones – southeastern England has neither metamorphic nor igneous rocks, as Figure 1 illustrates. In fact, Britain has few marbles, with those of South Devon and Iona being quarried only occasionally when stately homes demand. Most marble used in Britain is imported from Italy.

Whenever mudstones, clays and shales are subjected to the type of pressures described above their particles are flattened and their minerals turned round to follow completely new alignments. If they are also squeezed from the side the rocks will be re-folded and a totally new structure, known as 'slaty cleavage' imposed as the former sedimentary rocks are transformed into slates (see Figure 4). These break lines generally cut right across the almost obliterated bedding planes of the former sedimentary clayey rocks, but it is these new weaknesses that allow the slate quarryman to split the stone into wafer-thin sheets. It will later be shown how the slate mountains of Wales, Scotland and the Lake District have produced distinctive landscapes, sculptured not only by agents of erosion but also by the hand of man.

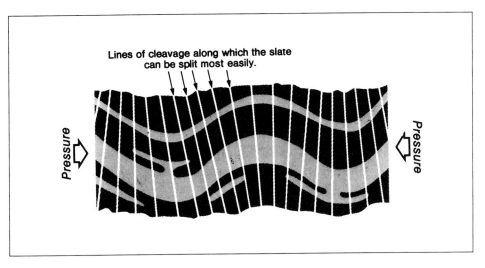

Lines of cleavage along which the slate can be split most easily.

Pressure

Pressure

Figure 4
The formation
of cleavage in
slate due to
lateral
squeezing.

On every occasion that Britain's crustal rocks have been tortured by episodes of mountain building, they have also experienced metamorphic changes over vast areas, often resulting from the collision of two ancient continents. Such gigantic pile-ups cause considerable crumpling, crushing and tearing of the crustal rocks, to say nothing of the associated volcanic activity and rock-melting at depth. It was on such occasions that the majority of Britain's metamorphic rocks were formed, especially those from which most of Scotland's Highlands and islands have subsequently been carved (see Figure 1).

# THE SCENERY EVOLVES

When the turmoil of mountain building had ceased some 15 million years ago, the enveloping oceans had withdrawn and Britain's landmass had been finally uplifted above the 'azure main', it must have presented a scene quite different from that of today. It is true that the uplands of Wales, northern England and Scotland would have projected their rocky summits steadfastly against the Atlantic winds and rain, sheltering, as they do today, the rolling plains of lowland England. But until two million years ago those mountainlands would have been relatively smooth, dome-like features not yet blemished by the vagaries of frost and ice. At that time the climate was sub-tropical and the entire area would have been clothed with a vegetation more akin to Florida than to Folkestone. Through the lush forests of exotic trees rivers flowed sluggishly to the sea adding their burden of sandy and muddy sediments to the swampy bays and estuaries.

The coastline two or three million years ago was also very different from that of today. Britain was still linked to the Continent where the Straits of Dover now intervene; East Anglia was no more than a wilderness of sand dunes and shelly coastal mudbanks crossed by an early river Thames flowing north-eastwards to debouch near modern Felixstowe into a shallow and minuscule North Sea. Farther south the Channel waves had only just succeeded in breaching the narrow chalk ridge which formerly joined the Isle of Wight to the Purbeck Hills of Dorset. The Isles of Scilly were still part of the Cornish mainland and the Welsh coast lay considerably farther westwards, with Anglesey remaining attached to the Welsh massif. The modern gulfs of Liverpool and Morecambe Bays were yet to appear, as were the firths of Solway and Clyde. The ragged fringes of Scotland's Western Isles were then for the most part an integral hem of the mainland fabric, as were the isles of Orkney. The eastern coasts of both Scotland and northern England projected some considerable distance beyond their present shorelines.

Because most of the high land lay in the west the majority of Britain's rivers flowed eastwards, just as they do today. The modern west-flowing rivers of the Severn, Mersey, Ribble and Clyde had yet to be born. Some writers believe that the Thames rose in Wales and that the Dee flowed out to the Wash long before Severn and Trent dismembered them. Whatever the case the unarguable fact remains that it was the rivers that continued to fashion the newly emerged landmass as unerringly and incessantly as they had done on previous landmasses for countless millions of years. It was the constant fretting of the pebbles against every river bank that caused undercutting and collapse, thereby ensuring that loose material was continually being carried downstream and that the river's 'tools' were always being replenished. In times of storm and of flood enormous quantities of soil and stones would have been washed from the hillsides, while landslides would have furrowed the steeper slopes. Multiply this seasonal degree of excavation by several million years and it is not difficult to see why the term 'everlasting hills' is nothing but a misnomer. The mountains of Wales and Scotland were once as lofty as the youthful Alps, but denudation has reduced them to a shadow of their original height.

Not only has running water helped lower the general level of the uplands but it has also played a paramount role in the fashioning of the lowlands. It has been seen how water whittles away slowly but surely at any weakness in a rock, enlarging its joints and loosening blocks of stone which finally topple down the hillsides. Such weathering is just as significant on the gently inclined lowland ridges as it is on the contorted rocks of the mountains. This explains why the exposed edges of the chalk downs and Cotswold limestone ridge (commonly referred to as 'escarpments' or 'scarps') are being nibbled slowly back, exposing the clays beneath to the ravages of erosion. Rivers have been able to cut down more rapidly into those less resistant rocks, creating broad clay vales and introducing lengthy lowland corridors beneath the stony-faced escarpments. The scenery of south-east England is little more than a succession of such scarps and vales: the Vales of Berkeley and Evesham have been incised below the Cotswold scarp; the Vale of the White Horse and the Vale of Aylesbury extend beneath the Berkshire Downs and the Chilterns respectively; the Wealden clay vales of Kent and Sussex are overlooked by the abrupt lines of the North and South Downs; farther north the Vale of Trent is shadowed by the ribby limestone scarp of Lincoln Edge. It is noteworthy that all these examples of scarp-and-vale land-scapes occur in lowland England; scarplands are less extensive in Britain's upland zone. Many of those in Scotland, for example, mark the edges of ancient lava flows while others both there and in the Pennines have been fashioned where major cracks in the earth's crust, known as 'faults', have caused whole tracts of stony upland to collapse along so-called 'fault-line scarps'.

Nevertheless, the junction between hard and soft rocks is not always marked by a cliff, escarpment or sudden fall of ground, notwithstanding the general rule of thumb that the harder the rock the higher the land. In reality it is often difficult for the layman to know when he is crossing from one rock type to another. There

are two major reasons why some geological boundaries have become progressively blurred or buried. First, it has already been explained how soil and pieces
of loose stone move slowly downhill. The effect has been to smooth out rocky
angularities on the slopes, to create gentler gradients and produce the hump-
backed summits and rolling hills and plains which characterize Britain. From the
majority of viewpoints an observer will soon realize that gentle curves far
outweigh the vertical and horizontal lines in the British countryside. Not for us
the angularity or the stature of the Grand Canyon or the Matterhorn. Indeed,
there were very few vertical edges in Britain's landscape until two million years
ago; then came the cataclysm of the Great Ice Age – the second reason why the
geological boundaries have been partly obscured. Geologists have in fact
produced two different types of geology map to illustrate the variations in
Britain's fabric. The first is known as a solid map since it depicts the characteristics of the solid rocks, generalized in Figure 1; the other shows the unconsolidated superficial materials such as peat, blown sand and above all the chaotic
clutter of boulder-studded clays, sands and gravels left behind by the Ice Age
glaciers. The latter deposits have been given the collective title of 'glacial drift'
since early geologists believed that the exotic stones must have drifted far from
their sources on ice floes. Thus the second of their definitive maps (see Figure 5)
is known as a drift map. A comparison of the two maps reveals that the solid
rocks of eastern and northern England have been largely blanketed with glacial
drift. But since the ice sheets never extended south of a line drawn from North
Devon to Essex the rocks of southern England were never buried by glacial
material. Nor were the majority of the uplands, for they were the source of much
of the drift. Before the onset of the Ice Age, rivers had played the dominant role
in fashioning the landscape; once the climate had deteriorated, however,
streams of ice became the major eroding and transporting agents.

As the refrigeration set in, Atlantic winds, which previously had supplied the
rain, now brought heavy snowfall to Britain's uplands. Drifting snow, blown off
the rounded summits by the prevailing south-westerlies, settled in the north-
east-facing hollows where, protected from sporadic sunshine, it gradually har-
dened into glacier ice. Each tiny glacier began to enlarge its hollow by removing
frost-shattered rock from its encircling slope, finally producing an armchair-
shaped void on the previously smooth mountainside. In Scotland such glacial
hollows are termed 'corries', in Wales 'cwms', and many of them now harbour
tiny jewel-like lakes or tarns where glaciers once were spawned. As the glaciers
expanded they advanced down the hillsides and eventually filled the valleys
previously occupied only by the rivers. Ice by itself is incapable of erosion but
supply it with numerous hard stones and it will cut down like a diamond saw

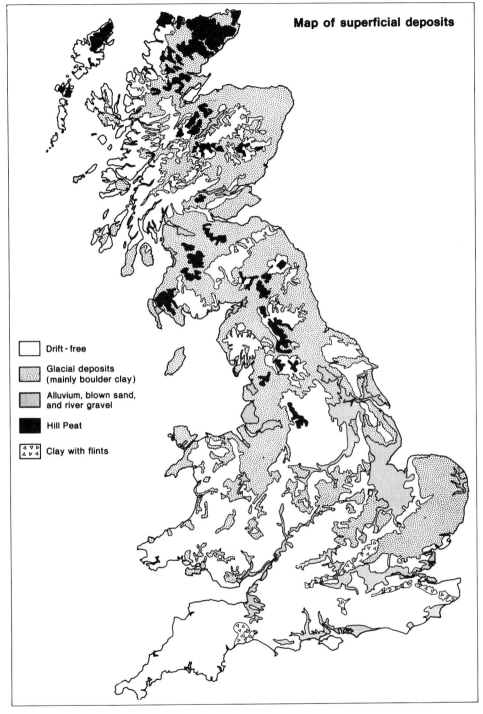

**Map of superficial deposits**

Figure 5

Drift - free

Glacial deposits
(mainly boulder clay)

Alluvium, blown sand,
and river gravel

Hill Peat

Clay with flints

The Nantlle
valley,
Snowdonia,
showing a
glacier-
moulded
hollow

through any kind of rock. Chunks of stone, split by frost from the mountainside, became the glacier's tools where, frozen into the sole of the ice, they were dragged abrasively across the rocky countryside, scouring, rasping and planing the uplands until they were completely remodelled. Moreover, frost was shattering the well-jointed rock outcrops above the glaciers till they rose like towers and minarets from the mosque-like domes of the dusky mountain skyline. As the corries were bitten savagely back into these frost-scarred peaks the intervening ridges became narrower and narrower until they were little more than serrated knife-edges, known as *aretes* (see Figure 6). Gradual coalescence of the neighbouring corries chiselled some of the peaks into pyramidal shapes, well exemplified by Snowdon or the Cuillins of Skye.

Prior to the Ice Age, rivers had carved their upland valleys into a series of curving and swinging corridors as their running waters meandered to avoid hard-rock obstacles. The glaciers had no such scruples as they bulldozed down the highland valleys, slicing off the ends of the ridges to form 'truncated spurs' (see Figure 6). In this way valleys were not only straightened but also excessively deepened into characteristic U-shaped forms. Even the toughest bands of rock were overridden and ground down into polished and grooved knolls of naked stone, termed 'roches moutonées' because of their resemblance to recumbent sheep. In a few thousand years Britain's uplands had been utterly transformed. Their gentle, curving outlines had been sawed, hacked and bludgeoned into a gigantic fretwork of splintered peaks, deeply chiselled gorges and razor-sharp ridges. Henceforth the vertical planes of precipitous cliffs and crater-like hollows

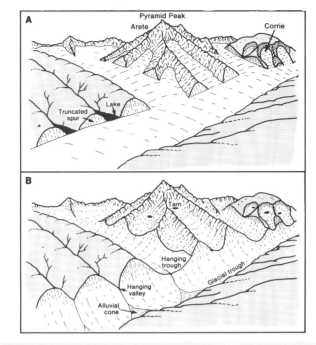

Figure 6 The formation of
glacial scenery in a mountain
landscape.
(A) During the Ice Age.
(B) Present-day scenery.

Honister Pass in the Lake
District. A typically glacially
eroded U-shaped valley.

were to dominate the scenery of Scotland, Wales and the Lake District. Today, these lonely places are mainly landscapes of stone, muted in a few favoured spots by pockets of soil and the softening effects of bog, river and woodland; but in general so recent was this latest chapter of earth history that nature has had insufficient time to heal the scars. However, it is time to give some thought to the lowlands of Britain and to wonder where all the stony material that had been removed by glaciers from the western mountainlands ended up.

It is difficult to visualize the ice caps as having been thick enough to bury even the highest mountain summits. It may be even more difficult for city dwellers in Nottingham and Norwich, for example, who are so divorced from the ice-ravaged uplands, to appreciate that thick ice sheets once prevailed where city streets and dwelling houses now innocently stand. But the thick, buttery glacial clays which were smeared liberally across the lowland zone have regularly been dug to supply the bricks for many a British city. It was these clays, studded with far-travelled mountain stones (hence the term 'boulder clays') which choked the courses of the lowland rivers and diverted the Dee, Trent and Clyde to their present courses. The melting ice sheets also produced such copious amounts of water that vast quantities of sandy gravels were washed out of the glacial debris and were spread far and wide, infilling the river valleys and festooning the hillsides with irregular hummocks and hollows. By carving the valley gravels into terraces and mounds, rivers providentially created excellent locations for future villages and bridging points, raised just sufficiently above the flood level. When Britain's urban expansion began in the nineteenth century these sands and gravels were to prove an invaluable source of construction materials. The geological drift map (Figure 5) illustrates not only the wide extent of the glacial clays and sandy gravels but also the fundamental dichotomy between the naked uplands and the partly draped lowlands.

The scene was now set for the arrival of Britain's earliest settlers, and what an obstacle course it presented. Forests and thick undergrowth had already spread along the clay vales and across the plains, flourishing in the post-glacial warmth and on the thick mantle of lowland soils. The ancient Briton faced a dilemma, on the one hand the lowland valleys would have been choked with impenetrable vegetation and waterlogged boulder clays too marshy to cross. In the uplands, on the other hand, the glaciers had swept away the soils so there he would have been presented with a stony waste or peaty moor, except in a few favoured places. While they remained merely hunters and gatherers, or even when they learned to domesticate livestock, the earliest settlers would have coped more easily with the bare stony hills and coastal headlands of Britain's western shores and this is where they chose to live. The New Stone Age peoples were the first to use lumps of surface stone to construct their rudimentary homesteads and their artifacts, for earlier Old Stone Age arrivals had utilized the natural nooks, crannies and caves of the western and northern fringes to provide for their domestic needs.

When the Neolithic farmers arrived in Britain they brought revolutionary techniques which included the introduction of primitive arable farming. Stony uplands and waterlogged lowlands would have been no use to them, for their type of cultivation required well drained nutritious soils. Not surprisingly, these New Stone Age immigrants found that the rolling grassy chalklands and the freely draining limestone scarplands of south-eastern England fitted their requirements perfectly, especially when the readily available supplies of flint more than compensated for the lack of other hard rocks, at that time essential for fashioning their tools and weapons. Thus, they built Britain's first real villages on the pale chalky hills and traded the flints by extending their remarkable Ridgeways far across the broad-shouldered downlands. These ancient routes can be regarded, in a sense, as Stone Age and Bronze Age 'motorways', never to be emulated until after the Roman invasions some three thousand years later. Henceforth, the destiny of the landscapes of stone was to be shaped increasingly by man.

From these uncertain beginnings the British began to fabricate their cultural mosaics upon the stony base. The earliest chapters in Part II of this book will examine how the first intrepid tribes tried to tame the bleak granite moors and rugged igneous hills throughout the length of western Britain from Cornwall to Orkney, lands of legend and misty wilderness, romantically referred to as the Celtic fringe. These are lonely places now; their population has dwindled and their quarries have all but closed. Nevertheless the rocky western hills and promontories can boast a remarkable record of stone production stretching back for 5,000 years. As man's technology became more sophisticated, he was able to build his settlements on the lowlands where trade demanded that estuary ports and bridgehead towns must be more accessible and strategically placed. By and large the lowlands are synonymous with the sedimentary rocks although it has been shown that large tracts are blanketed by drift. But how much more effective were those towns which could raise their castles on protruding crags, like Edinburgh and Nottingham. In the troubled years of Britain's middle age a commanding and defensible site was everything. Not surprisingly, therefore, on the flat river plains where solid rock was irredeemably buried, the Normans were forced to throw up artificial mounds ('mottes') of earth as their defensive points, unless they were fortunate enough to discover an accommodating rocky eminence like the isolated chalk knob at Windsor Castle, right alongside the Thames. The later chapters of Part II will describe more fully how the sedimentary rocks have contributed to the scenery. It will be seen how the ridges of limestone and sandstone were to become the greatest source of Britain's building stone for almost one thousand years and why, for example, the Pennine gritstone moors have a character quite different from that of the flint-strewn chalklands. The concluding chapter of Part II will describe how the Victorians' demand for slate to roof their mushrooming cities led to a remarkable boom in some settlements of the western hills and mountains.

# THE LANDSCAPES OF BRITAIN

# THE GRANITE LANDS

Granite is synonymous with strength, hardness and durability, characteristics which make it one of the toughest of natural building stones. This massive, coarse-grained, noble stone epitomized all the virtues of Victorian England, where it became a symbol of endurance and almost of morality. Except in the south-west, where it was the local stone, granite was rarely used in English building until Victorian times but then, because of improved quarrying techniques, its bold geometric lines could be found in almost all of the new public buildings – Jacquetta Hawkes spoke of it as the 'pillar of Victorianism'. Nonetheless, a town constructed wholly of granite may look sturdy but can also be uniformly dull, except after rain when the mica sparkles. Such is the belief of Alec Clifton-Taylor who lists the reasons for granite's apparent monotony as a building stone: first, its colour and texture never change with weathering, making it impossible to acquire the patina of a sedimentary freestone; second, unless it is polished it fails to reflect the light as well as a sandstone or limestone; third, it does not yield easily to the chisel so that its lack of ornamentation renders it incapable of throwing crisp shadows.

Such critical feelings may certainly be aroused when appraising the contributions made by granite stone to the modern British townscape, but when its character is judged only in its natural rural setting then it may well evoke quite different responses. A visitor to most of Britain's granite moorlands will probably agree that the homely churches and the sturdy cottages of native stone sit easily in the landscape where they speak more eloquently than any history book. The rough-textured, unpolished granite blocks, especially where lichens have taken hold, give a comfortable solidity to the manors, farmhouses and barns, their mellow character matching both the colour and the texture of the surrounding moorlands whence they came. Many of these rural granitic buildings look as if they have grown out of the ground, as indeed some of them have, for they had to be raised on the spot wherever the surface blocks were too massive to be moved. Perhaps more than most landscapes of stone, granite lands exhibit a remarkable affinity between man and environment, possibly because the solemn contours of the intractable stone evoke visions of a primeval relationship that has scarcely changed. A visit to the mist-shrouded standing stones of Dartmoor or Penwith will confirm such feelings.

Lower Tor farm, Poundsgate, Dartmoor. A granite-built farm founded on moorstones too large to move.

One might be forgiven for thinking that granite's apparently indestructible qualities must identify it with Britain's highest hills and mountains. It is true that in most instances its resistance to weathering has made the granite bastions the reliable foundations around and upon which the superstructure of Britain's other highland rocks has been firmly pinned. But, like any other rock, granite is affected by rain, frost and wind, despite its reluctance to yield to nature's bludgeoning.

When viewed from afar, whether highland or lowland, the landscapes produced by granite are generally smooth and only rarely do they verge on the spectacular. Once their cover of layered rocks is removed, the previously buried igneous domes almost invariably weather into smooth-shouldered hills or whale-backed mountains. Because these massive rocks have no bedding planes, they offer relatively few chinks for the elements to exploit. Nevertheless, there are sufficient joint weaknesses to undermine the strength of even this most stubborn of rocks. Once the rock's smooth surface becomes pitted tiny pools of water accumulate and these accelerate the rotting until bosses of rock become isolated from each other and surrounded by aprons of sandy gravel known locally as 'growan'.

On the larger granite domes, where joints are few and far between, the rock breaks down in a different way. Over thousands of years thick curvilinear sheets of granite are shed periodically from the dome, as if an onion is being peeled of its outer skin. Eventually the sheets themselves disintegrate into a scatter of granite blocks around the bald-headed summits and these are the so-called 'moor-stones', used for primitive construction in these intractable granite lands. Many a tourist has discovered that the apparently smooth outlines of the granite hills prove to be disconcertingly rough when approached on foot, especially where the confusion of broken blocks is marked by heather and bracken.

It is the more detailed natural sculpting of granite that is best remembered by the visitor: the castellated tors of Dartmoor and the Cairngorms or the pinnacled cliffs and chasm-like zawns of Cornwall's farthest coast. The architectural character of tors led our ancestors to see them as works of the Devil or of legendary giants. Close inspection certainly suggests that their titanic 'masonry' was artificial. The truth is rather more prosaic. Two schools of thought have contrasting views on tor formation. One believes that these battlemented rock features were spawned deep beneath a tropical land surface several million years ago when warm percolating rainfall rotted the bedrock to depths of some 50 feet. The latticework of joints allowed the rotting to etch out rough, blocky outlines in the buried granite in the same way that a virgin stone is provisionally blocked out by a sculptor. While the groundwaters were dismantling the granite at depth, atmospheric processes were also at work lowering the covering rocks until, after millions of years, the buried tor began to emerge at the surface. Near Two Bridges on Dartmoor a partly emerged tor can be seen in an abandoned quarry. Finally, frost, wind and rain added the finishing touches to leave the turreted rocks towering above a boulder-strewn platform. Indeed the word 'tor' is derived from the Celtic '*twr*', meaning tower. An alternative view believes that such monolithic forms could never have been produced furtively within the humid stygian gloom of a decaying underworld but that tors must have been boldly hacked out in the crisp frigid daylight of the Ice Age. This view suggests that severe frosts would have chiselled freely away at any well jointed mass of rock, no matter whether it was unbedded granite or bedded sandstone. It can only be concluded, therefore, that tors are not the preserve of the granite lands, although they are seen most commonly in granitic terrains.

# Cornwall

This knuckly peninsula, projecting bravely into the Atlantic swell, exhibits some of Britain's most awesome granitic scenery. Whether it be the mist-veiled tors of Bodmin Moor, the industrially scarred tracts of St Austell and Carn Menellis or the rugged coast of Penwith, there is no mistaking the uniqueness of the Cornish landscape. Its rain-washed oceanic light (which inspired the artists' colonies of

Newlyn and St Ives) softens the austere rocky vistas; the encompassing sea frets unceasingly at its beleagured coastline; its heathery moors, littered with lichen-crusted stones, cradle remnants of long forgotten cultures. Such are the basic ingredients of Cornwall, a land where primeval granite dominates the human imprint as surely as it pervades the natural scenery. Cornwall's granitic tracts have weathered not into a soft and ordered landscape of manicured fields and hedgerow trees but into one of bleak severity whose veneer of being tamed comes from centuries of toil in stone-choked fields, on wave-battered coasts and in gloomy quarries. One quickly gains an impression that early man's life here must have been as hard as the granite itself.

The majority of summer visitors will carry away memories of the sunlit colour-washed stonework of quaint fishing villages, the historic grey ruins of the tin mines, the aquamarine waters of the sandy coves and the pinnacled granitic splendours of Land's End – the Cornwall of the postcard and calendar. But if one cares to look beyond this colourful façade, the landscape has an even more fascinating story to reveal. Venture up on to the moorlands or walk the coastal path and you soon become aware of the true nature of the granite lands. Anyone who has experienced the loneliness of Bodmin Moor or stood facing an Atlantic gale at Cape Cornwall will recapture something of the environment encountered by the earliest Cornish settlers. These are places where starkness prevails, where hamlets crouch in folds of the land or fishing villages huddle under the sea cliffs, sheltered from winter gales that few visitors experience. It is little wonder that the roofing slates have been slurried with mortar or smeared with tar to keep them wind and water proof; for the same reasons the chimneys of the granite cottages are broad and squat, their windows are small and their neighbouring church towers short and stumpy. The traditional architecture reflects not only the difficulty of the almost treeless, wind-seared terrain but also the unyielding nature of the only available building stone. Until the nineteenth century granite was too hard to quarry economically, and the population was forced to make widespread use of the ubiquitous moorstones which are irregular in size and shape. The larger boulders were always used at the base of the cottage walls, the smaller ones placed indiscriminately above, the squarish ones reserved as corner stones, and the longest used exclusively as lintels for doors and windows. In the church of Altarnun on Bodmin Moor, for example, moorstones have been used not only as the footings of the walls but also for its primitive pillars.

Although not as extensive as Dartmoor, Cornwall's Bodmin Moor has all the character of a wild granite landscape, with such grizzled summits as Brown Willy, Rough Tor, and Cheesewring encapsulating the special flavour of the Cornish scene. Their battlemented tors crown the heathery moors across which '. . .

Lanyon Quoit, Penwith, Cornwall. Such mighty granite monuments, constructed by Stone Age man, inspired Barbara Hepworth's sculpture.

unfenced roads wind between granite boulders and isolated cottages, with patches of treacherous marsh between'. Thus wrote John Betjeman who knew his Cornwall better than most. His perceptive eye described how '. . . ancient granite "hedges", going back to the Iron Age, curl round brackeny fields on the moor', where 'ferny, high-hedged lanes dip down to the brown river Fowey'; how other moorland streams flowed in '. . . deep wooded and winding valleys crossed by packhorse bridges and old stone-arched bridges'. He appreciated how early Cornish men used granite boulders to build the delightfully named Bronze-Age villages of Stripple Stones and Trippet Stones on the moor and why the Celts laboured mightily to clear the moorstones from their tiny fields. Betjeman equally appreciated the later buildings, especially the granite church of Blisland – '. . . the most beautiful of all country churches of the West'. Blisland is built of a brownish-grey granite from the nearby De Lank quarries below Rough Tor, the source from which the carefully dressed blocks were selected for the famous

The prehistoric stone circle of the Hurlers, Cornwall, with a derelict tin mine beyond.

Eddystone lighthouse. To the east of the moor, below Betjeman's 'gigantic nodding mushrooms' (as he called the Cheesewring Tor) the ruined engine houses and the gaunt granite cottages are a reminder that the Caradon copper mines once flourished here. These, together with Trevethy's prehistoric cromlech, the Bronze Age stone circles of The Hurlers, the granite churches of St Cleer and North Hill, are simply further pieces in the stony jigsaw of this primordial scene. North Hill's cottages may be slate-hung but their exterior granite staircases echo the solidity of their hidden walls. Not only has the intruding granitic mass baked the surrounding shales into slates but it also created the dark greenish metamorphic stone known as polyphant, from which Leewannick's fifteenth-century church was built.

The peninsula of Penwith, the very toe of Cornwall, has a different type of granite landscape. Whereas Bodmin climbs up to the mists, Penwith crouches down by the shore. Virtually surrounded by the throbbing Atlantic, the ridge of low heathery hills between St Ives and Land's End has been aptly described by Patrick Heron, the art critic and painter, as a landscape dominated by '. . . the bleak, bony, strong mysteriousness of the Celtic Moors'. Penwith is certainly the most Celtic part of England and that most akin to Brittany; like the latter it has a profusion of prehistoric stone monuments and ruins. Its standing stones (locally known as 'menhirs'), its circles and carved crosses, and above all the cyclopean boulders of its burial chambers ('quoits') mimic the natural architecture of the monolithic granite tors and their occasional delicately balanced rocking stones ('logans'). It is not difficult to see why Barbara Hepworth, in neighbouring St Ives, was inspired by this singular countryside to create some of her most memorable sculptures. On the treeless, boulder-strewn moors behind Zennor the forlorn remnants of Chysauster prehistoric village have been likened to those of Shetland's famous Jarlshof, while the hoary stones of Chun Hill's Iron Age fort and village in Morvah parish are equally celebrated. Archaeologists have excavated no less than twenty-three such prehistoric villages in Penwith alone, each granitic cluster of courtyard houses surrounded by a hotch-potch of ploughed Celtic fields.

If the moors are haunted by the spirits of these primitive Cornish farmers, then the coasts are redolent with more recent ghosts of miners and fisherfolk. Perched beneath grey, fretted cliffs or tucked into sheltered coves, the ruined tin mines and the sturdy fishing quays testify to the once thriving but certainly tough lives of working-people on this rock-bound coast, a way of life immortalized by Stanhope-Forbes and his Newlyn school of artists. Tin was only one of the metals produced by the mineralizing vapours rising from the granitic mass, but it was tin that drew the Mediterranean invaders from Carthage and Phoenicia, centuries before Christ was born, once they had discovered that mixed with copper it formed a hard but malleable alloy known as bronze. Between the 4th and 14th centuries AD Cornwall's mineral wealth was almost forgotten, but by Tudor times, when Richard Carew was observing that '. . . the cliffs thereabouts muster long streaks of glittering hue, which import a show of copper', metal mining was again in full swing.

Today the gaunt chimney stacks and broken walls of the derelict workings are revered as ancient monuments at such places as the Botallack mine near Cape Cornwall. The miners' cottages, churches, farms and gnarled stone 'hedges' are all built from silvery-grey granite in such workaday villages as St Just and Pendeen. But their sombre tones are transformed at the coastline itself. The crenellated sea cliffs, emblazoned with yellow lichen, sea pink and thrift, are interrupted by coves of dazzling white sand where the granite has been pummelled into millions of crystals of sparkling quartz and spangled mica. The intricate

A derelict tin mine, Cornwall.

details of the Land's End sea cliffs serve to illustrate how this relentless ocean has exploited the weaknesses of the stubborn rock. In zones where cracks are closely spaced, shingle-charged waves have ground out the foam-drenched clefts, known as 'zawns', while such intervening buttresses as Dr Syntax's and Dr Johnson's Heads have survived where joints are fewer. Over the centuries their stony profiles will gradually weather away, just as the pinnacled sea stack of the Armed Knight is destined to lose its final battle with the tireless Atlantic. For the long scatter of half-drowned tors and reefs, stretching out to Longships light-house, the unequal war of attrition is almost over.

In the picturesque fishing villages of Sennen, Lamorna Cove and Mousehole, the greenish-grey granite of the cottages and quays has today been brightened by pastel shades of colour wash. In addition the hard, geometric lines of their stone walls and chimneys have been broken by a riot of fuchsias, hydrangeas and geraniums. The granite walls and pavements of such towns as Penzance, Penryn and Helston however, have not been muted to the same degree. In urban situations, where the blocks were cut to size and carefully laid in regular courses with closely fitting joints, one feels that granite has forfeited much of its romantic

Land's End, Cornwall, with the pinnacled sea stack of the Armed Knight behind the arched isle of Enys Dodnan.

appeal. Paradoxically, once the techniques for granite-working became firmly established, Cornish architecture lost much of its robust charm.

# Dartmoor

This well known Devonshire moorland represents the largest granite massif south of the Scottish Highlands and its highest summit, High Willhays (2,038 ft), the southernmost bastion of upland Britain. In contrast to those of Cornwall, the Devonshire granite lands fail to reach the coast, yet their magnificent scenery

Bowerman's Nose, a granite tor on Dartmoor.

ensured that in 1951 Dartmoor became the fourth of Britain's National Parks. Its sweeping tableland, rising dramatically above the rich Devon farmlands, which Gerard Manley Hopkins described as the 'soft maroon or rosy-cocoa-dust-coloured handkerchiefs of ploughed fields', is little more than a heathery wilderness of bracken and wind-flattened grass. The treeless miry hollows alternate with gently swelling tor-crowned hills. Nothing breaks the stillness of its lonely valleys save the croak of the ravens and the bleat of the mist-dampened sheep, until sporadic storms charge across the moor; then it becomes an uncompromising place. Its slowly running brooks are rapidly transformed into raging torrents, tearing at the tussocky banks and carrying quantities of soil and pebbles swiftly down their valleys. Once the peaty ground is stripped away, the bleached granitic rocks become exposed to the rigours of frost and rain, perhaps destined to be the tors of future millennia.

The different character exhibited by each of Dartmoor's tors can be explained by their contrasting joint patterns. The impressive tower-like forms of Hound Tor, Bowerman's Nose and Haytor East, for example, result from the predominantly vertical cracks formed when the granitic mass was first pushed up. Where the jointing is primarily horizontal, however, the layering (similar to a bedded sedimentary rock) only appeared when the granitic surface 'expanded', once relieved of its burden of overlying rocks. In this case the resulting rock piles are squat and more tabular, like those of Bellever Tor, Honeybag Tor, and Yar Tor.

In prehistoric times these Devonshire moorlands, together with the neighbouring Cornish promontories, housed an early British population whose numbers were matched only by those of the Wessex chalklands. Yet Dartmoor has yielded surprisingly few pottery sherds or metal objects, its earliest artifacts being limited to a few flint arrowheads and stone axes. By the Bronze Age (1800–500 BC), however, the inhabitants of the moor had made considerable use of the scattered moorstones to construct their huts, pounds, circles and 'avenues' of standing stones. From the surviving patterns of these granite boulders it is possible to reconstruct something of the cultural landscape that must have prevailed some three thousand years ago. Although the climate must then have been warmer and drier than at present, the earliest farmers built the majority of their villages and livestock enclosures (pounds) on the southern half of the moor, suggesting that even then the higher, boggy northern tracts offered little attraction to man or beast. Grimspound is the best known and most impressive of the enclosures, comprising a massive but ruined ten-foot thick retaining wall which encloses the remains of twenty-four hut circles. Here the villagers would have protected their cattle and sheep from marauding bears and wolves, although it is less easy to explain the purpose of the henge-like circles or stone avenues except in the context of some pagan ritual.

By the Iron Age (4th century BC), when British weather had deteriorated, the moor had been virtually abandoned. The large hill-forts of that time are found only on the peripheral foothills, and even the Romans rarely ventured across the heathery wastes. Until its tin deposits were actively worked from the mid-12th century AD, Dartmoor was to remain a wilderness for some sixteen hundred years.

For a few decades in late-Norman times Dartmoor became the major source of Europe's tin, with most of the early workings being opened in the south-west, near to the tiny granite-built village of Sheepstor. Although Devon's tin production had virtually ceased by AD 1300, just as the Cornish mines began to recover, the introduction of shaft-mining in the 15th century caused a renewal of the moor's tin industry, a bonanza which was to last for three hundred years. This period saw the expansion of Dartmoor's perimeter towns and the construction of the moorland villages' granite church towers, both paid for by profits from tin. Widecombe-in-the-Moor has the most notable tower whose grace and height have led to comparisons with a cathedral. Tin, having been washed out of the decayed granitic gravel, was smelted on the open moor in small granite structures known as blowing houses, where water wheels powered the furnace bellows. In the surviving examples, like that at Black Tor Falls near Princetown, one can still see the granitic blocks once used as moulds for the ingots. From these isolated

Sheepstor village, Dartmoor.

sites the metal was taken on pack horse to be weighed at one of the tin assay centres known as the Stannary towns of Tavistock, Ashburton, Plympton and Chagford. Of these only the latter is located on the granite outcrop – at a point where an ancient granite bridge crosses the river Teign. Below the bridge the attractive wooded valley is overlooked by the redoubtable Castle Drogo built by Lutyens between 1911 and 1930 and described by Pevsner as an 'extravaganza in granite'; it was the last major private dwelling in Britain to be built of granite.

More typical examples of Dartmoor's granite dwellings are to be found in such picturesque villages as Lustleigh or Buckland-in-the-Moor where W.G. Hoskins found that '. . . the prevailing tone of the human landscape is a warm grey; grey thatched roofs, grey moorstone walls and buildings, and tall grey beeches around them'. The trackways linking the scattered moorland villages were forced to cross the rivers by means of stepping stones or simple clapper bridges made from long slabs of moorstone, of which the best surviving example spans the East Dart near Postbridge. A few isolated manors and farms, some dating to the time of Elizabeth I, have survived on the open hills. Many of the farmhouses were built with one gable end hard against the hillside, with the house, byre and barn all capped by a continuous, smoothly thatched, gently dipping roof. Thus, as John and Jane Penoyre observed, '. . . the sloping roof ridge gives the Dartmoor farmhouse an indeterminacy of outline that merges it with the landscape in a manner curiously at variance with its sturdy granite walling'. It is hard to imagine how the yeoman farmers managed to wring a living from the seemingly barren moorlands but it must be remembered that it was centuries of grazing by domestic animals that produced Dartmoor's treeless vistas. Dartmoor must once have been forested, judging by the few patches of wizened primeval woodland, such as Wistman's Wood, that have precariously survived. By the time the open moorland was formally enclosed in the nineteenth century the natural woodland had all but disappeared and as the geometric fieldwalls were impressed upon its wilderness Dartmoor suffered a further indignity.

Large commercial granite quarries were opened at Haytor and Foggintor, producing smoothly dressed stone not only for the austere prison at Princetown (1806) but also for a myriad of public works throughout southern Britain. Dartmoor granite, impervious to water and air pollution, was utilized for example, in London Bridge (1823–31) and a number of docks, breakwaters and lighthouses, to say nothing of ubiquitous kerbstones and road setts.

# Scotland

There can be few greater scenic contrasts in Britain than between the low granite moorlands of South-West England and the high granite mountains of Scotland. The detailed sculpturing of the rock may be the same in Cornwall and Scotland, for both regions bear the hallmarks of granitic stone, but the scale of the Scottish mountains is so immense that if you could pile the four highest Cornish hills one atop the other, their combined height would still be less than the Cairngorms' loftiest peak. Granite plays a substantial part in the make-up of Scottish uplands and nowhere more than in the eastern Grampians. Here, the beautiful scenery of Royal Deeside is overlooked by the Cairngorms, Lochnagar and Mount Keen which together form a granitic backdrop unequalled in Britain. This is a land of ice-sculpted peaks, lonely lakes, spectacular waterfalls, swift-flowing, salmon-filled rivers and magnificent pine forests.

The Cairngorms can claim the most extensive area above 4,000 feet anywhere in the country, and their National Nature Reserve is the largest in Britain. Once known as Monadh Ruadh (Gaelic = Red Mountains), the Cairngorms have been carved from a gigantic mass of rose-coloured granite. Its dominant colour is derived from the reddish feldspar minerals although its flecks of mica and large quartz crystals (some of which form the yellowish Cairngorm gemstone) add to the rock's glistening sheen. It is when the water-smoothed granitic boulders and pebbles are seen gleaming in the riverbeds, however, that one can best appreciate both the crystalline beauty and the durability of the stone, qualities which explain why polished granite has played such a major role in the adornment of Britain's civic buildings. Nevertheless, although its toughness has also helped build some of Britain's highest peaks, the flat-topped summits of Lochnagar and the Cairngorms lack the dramatic profiles of many other Scottish peaks, except where their tors intrude like warts on the countenance of the plateau. Not surprisingly, some writers have suggested that these desolate, gently swelling granitic landscapes are lacking in picturesque appeal, although the majority would probably agree with W.A. Poucher, the well-known landscape photographer, that '. . . as the mountaineer wanders through and over them he will be captured by their remoteness and solitude, whilst the immensity of their scale will impress him with a wonder more profound than that experienced elsewhere in our island heritage'. No one would dispute the climatic severity of their windblown summits, where a sparse Arctic-Alpine flora clings tenaciously to the few pockets of stony soil in this frost-shattered wilderness. A few patches of scarlet creeping azalea also manage to survive in this gravelly 'desert', but even in high summer the vivid pink cushions of moss campion are often streaked with snow. The evidence of former glaciers is everywhere to be seen and where these have bitten deeply into the granite massif the gently rolling plateau is suddenly transformed. Its boulder-strewn summits are sharply truncated by ice-gouged precipices, ribbed with tor-like pinnacles and seamed with semi-permanent snow patches on the ice-smoothed granite slabs. Each of the mountains has its own suite of these glacially carved depressions, or corries, but few can compare with Byron's '. . . steep frowning glories of dark Lochnagar'.

Amidst the highest rocks of the Cairngorms' frost-riven landscape, where ptarmigan, dotterel and snow bunting nest between the lichen-crusted boulders, the infant river Dee is born as little more than a seepage from the plateau bog. After tumbling down the stony façades of these ageless domes the youthful stream wends down the enormous glacier-deepened trough of the Lairig Ghru, overshadowed by two of Scotland's highest peaks. As it takes leave of the bare mountains and seeks refuge in the juniper thickets and forests of Scots Pine its headlong surge through pink-walled gorges slows down to a more leisurely dawdle through gravel-choked meadowlands. In the fast-flowing stretches, grey wagtails, dippers and ring ouzels lurk among the water-splashed pebbles but where the current slackens the common sandpiper finds a home on the sandy river bank. On reaching the broader valley floor the Dee is forced to swing around huge granitic borders dumped by former glaciers, although its progression is now more stately as befits a river entering the Royal demesnes around Balmoral. On the flanks of the glens orange-trunked pines rise above a tangle of blaeberry, cranberry and heather, an ideal habitat for some of Britain's rarest creatures: capercaillie, crossbill and crested tit, red deer and red squirrel, pine marten and wild cat, although the last two take pains to avoid the haunts of man.

Despite the recent influx of skiers and tourists in the northern corries the Cairngorms have been little affected by man. Of prehistoric settlement there is no sign and, considering the harshness of this primeval landscape, it is little wonder that the region remained unpopulated until Victorian times. The few crofters who had eked out a living in the lower glens had to abandon their heather-thatched cottages in the 18th and 19th centuries, leaving nothing but a few broken fieldwalls and an occasional tumbled ruin to give shelter to both sheep and grouse. Thus rough granite buildings are rarely seen in the uplands, though carefully worked silvery granites brought from the coastlands may be seen in several of the Victorian churches, castles and hotels along the Dee valley. Unfortunately, the Caledonian pine forests which once clothed the granitelands from 'Glencoe eastwards to the Braes of Mar' have been seriously depleted by axe, fire and livestock, leaving the modern hills little more than a vast sheep farm.

Where the Dee eventually debouches into the North Sea a very different landscape prevails. Here, the lonely rubicund crags of the granite mountainlands are replaced by the bustling grey streets of Aberdeen with its broken skyline of towers, spires and multi-storey blocks. The basic character of this sturdy townscape is governed by the local silvery-grey granite which its citizens have laboriously prised from their own Rubislaw quarry since the mid-18th century. This local granite's blue-grey feldspars, though less eye-catching than the rosy crystals of its Cairngorm counterpart, have contributed to the dignified tone of the 'shining mail' of the so-called Granite City. Writers still romanticize about a 'glitter of mica at the windy corners' of Aberdeen's newly commercialized fabric. The city grew up by amalgamation of two early settlements, that of Old Aberdeen near the mouth of the Don and the fishing village of New Aberdeen on a tiny tributary of the Dee. Their original buildings were constructed from roughly squared granite blocks or the water-worn boulders picked from the riverbeds. Away from these historic rustic quarters, however, one can trace how the quarried stone becomes more finely worked, culminating in the remarkable light-grey façade of the Marischal College. In fact the latter was built from a different granite brought several miles from a Don-side quarry at Kemnay, whose stone was also utilized in the docks of Leith, Newcastle, Sunderland and Hull. Nonetheless, the greater part of Aberdeen has been constructed from the medium-grained silvery-grey Rubislaw stone whose quarry in the city suburbs has finally ceased production, leaving behind a 465-foot crater as the deepest

Black Rock Cottage at the edge of Rannoch Moor, where the granite has been worn down to form a basin.

quarry in Britain. Fittingly, one of its final uses was for the statue of Robert the Bruce erected at Bannockburn battlefield, but earlier its quarried blocks had helped build the docks at Sheerness, Portsmouth and Southamptom, to say nothing of the Bell Rock lighthouse. Solid granite remained the major building material until the end of the 19th century, but because of rising costs many of the later houses of Aberdeen's inner suburbs are merely granite-faced. Finally, as costs became prohibitive and granite quarries closed, the city's outer fringes have been built of non-granitic materials, just as clay tiles have replaced slates as a roofing material.

To conclude the picture of Britain's most famous granitelands something has to be said of the spongy, streaming wilderness of Scotland's Rannoch Moor, which in R.L. Stevenson's *Kidnapped* was shown to be a trackless and weary-looking desert. Its desolate but awesome landscape is a waste of dun-coloured grass, flecked with grey rock, blotched with black peat and dotted with pools of sullen water. Yet the surprising aspect of this extensive blanket bog is the fact that the great amphitheatre of the moor is floored by granite and surrounded on all sides by towering non-granitic peaks. If granite is one of the most resistant rocks one has to explain why the Rannoch granite coincides with a basin, especially since granites of similar age build the Cairngorms. The answer lies partly in the toughness of the perimeter rocks which include very hard quartzites and volcanics, many of them baked even harder by the heat of the invading granite mass. Thus, rivers have fashioned a shallow basin and subsequent ice sheets have scoured out the thick accumulation of rotted rock and sour soil. The laws of nature seem to have been flouted on this occasion.

# LANDSCAPES OF LAVA

Many landscapes of stone possess a gentleness of form in which rolling hill and dale or undulating countryside are undisturbed by jagged rock formations. Such uniformity of shape strongly suggests that the underlying rock structures have such similar properties that they are weathered away at almost equal rates. The tranquil English chalklands are a good example of this scenic blandness while even some granite moorlands possess a smoothness of profile that belies the wizened sturdiness of their rocks. Most volcanic landscapes, however, exhibit a tortured look, there is nothing bland about a recently formed lava flow with its jagged black clinker and blanket of searing ash. Even after countless years of weathering, in which the original volcanic cone may have been completely destroyed, lava terrains tend to be bold, stark and somewhat forbidding, particularly in colder regions where glaciers have scoured away their coating of soils. Yet their broken skylines and rugged shapes often possess a majestic grandeur wherever they project their irregular boldness into the symmetrical uniformity of the surrounding sedimentary rocks, as if to mirror something of the violence which accompanied the formation of the volcanic rocks. Above all, many of their stones are satanically dark, redolent of the underworld whence they came. Volcanoes are, of course, surface manifestations of this fiery under-world, their vents being little more than chimneys of the subterranean furnace in which all types of crustal rocks are melted down into a white-hot seething fluid known as 'magma'.

Today, Italy possesses all three of Europe's active volcanoes and Britain has no volcanic activity of any kind. But an examination of geology maps will reveal that periodically throughout Britain's geological history our tranquil scene has been punctuated by spectacular eruptions which produced blankets of volcanic ash and extensive spreads of surface lava to say nothing of the magma which squeezed between the crustal rocks to create sills and dykes. Many of Britain's most spectacular escarpments have been sculptured from these sheets of black intrusive rock or from the thick piles of dark lava. Furthermore, the well-known mountain landscapes of Snowdonia and the Lake District owe much of their ruggedness to the angularity produced by their tough volcanic rocks. In both of these cases flows of lava and bands of volcanic ash became interbedded with

layers of sandstones and shales, for when the volcanoes exploded almost 500 million years ago they erupted not on the land but among the sandy and muddy sediments of the sea floor. Finally, when all volcanic activity ceased and the mighty cones had been worn down to mere stumps, it was the solidified lava in the vents that proved to be the most resistant. Thus the sporadic lava 'necks' were left as prominent isolated hills in such places as Edinburgh and the Scottish lowlands where they have provided valuable craggy citadels for many an early settlement (see Figure 2).

Like all other igneous rocks, those of volcanic origin have a crystalline structure but because the volcanic crystals are so tiny and their colours so muted they pale into insignificance alongside the gem-like lustre of granitic stone. Only in the few instances where yellow or pink lavas are found, as in parts of Snowdonia, for example, does volcanic stone depart from its sombre brown, black or grey colouring. This may be explained by its chemical composition, for most British lavas, such as basalt or dolerite, have a high proportion of dark-coloured iron minerals which, when weathered, give the rock surface a veneer of 'rustiness' which merely adds to its drab appearance. Little wonder

Nant Gwynant, Snowdonia. A landscape carved from ancient lavas.

that most volcanic rocks have never been in great demand for building stones except in areas where no other suitable materials were to hand. Even in the Lake District and the Hebrides, where volcanic rocks have been widely used in the vernacular architecture, the ancient cottages are often liberally coated with whitewash to brighten up the sullen stone. Nonetheless, what volcanic stone lacks in visual appeal is more than compensated by its hardness and durability and wherever it is used as a building stone the masonry has an enduring look about it. Moreover, many British roads are constantly being resurfaced with 'chippings' of crushed basalt or dolerite which stand up to the pounding of heavy traffic far more successfully than any of the widespread and more easily acquired sedimentary rocks.

## The Lake District

The entire central massif of the English Lake District has been carved from a variety of very old volcanic rocks which themselves have been folded and bent. This rugged tract boasts England's highest summit, Scafell, but it is not simply the height or the cragginess of its bare fells that gives Lakeland its remarkable appeal: its charm lies equally in the deeply scored valleys and their slender lakes. Few places in Britain can match such a treasurehouse of upland scenery ensconced in so compact an area. Within a few square miles there are notable scenic contrasts: the astringency of the stony summits and the pinnacled ridges is tempered by the burgeoning oakwoods of the vales; the wild abandon of the bouldery ghylls and their lonely waterfalls is offset by the intimate greystone hamlets compactly set in a network of tiny fields; the brooding, stone-framed tarns of the mountain hollows cascade quickly to the sparkling, reed-fringed lakes on the valley floor.

Much of this intricate detail of central Lakeland has been fashioned by former glaciers which quarried away the softer sedimentary rocks fairly easily but found the interbedded volcanics a tougher proposition. Thus, though all the fells have been ice-scraped and frost-shattered, it is never difficult to pick out the volcanic rock eminences – they are bony, gnarled and knobbly. Their very names echo their armour-plated, thrusting bulk: High Stile, Pillar, Steeple, Great Gable, Great End, Crinkle Crags, Pike O'Stickle, Steel Fell, Helm Crag, Nab Scar, Gavel Pike and Striding Edge, all familiar to the Lakeland tourist. By contrast, the northern and southern tracts of slates, shales and limestone, which flank the central volcanic belt, give rise to scenery that is altogether gentler, less broken and, with the major exception of Skiddaw, is markedly lower in elevation. Both the northern and southern fells are characterized by their more extensive woodlands and their greater spreads of heather for there the soils are less sour, the rainfall less heavy, the hills less elevated and the slopes less precipitous. One only has to travel southwards from the steeply confined defiles of Langdale and Rydal Water to the broader extravagances of Lake Windermere to witness the change of scenery. Equally dramatic contrasts will be seen by the traveller leaving the overwhelming volcanic terrain of Borrowdale and journeying north past the placid expanses of Derwentwater and Bassenthwaite en route for Cockermouth. These northern and southern lake shores are edged with woodland and backed by moderate slaty hillslopes patched with bracken, heather and stone-walled plots of grazing land – bare rock outcrops are relatively uncommon. To experience the savage face of Lakeland one must visit the wild and primitive Wasdale whose silvery Wastwater is surrounded by beetling crags and overshadowed by towering peaks of volcanic rock. When early visitors spoke of the lakes 'as beauty lying in the lap of Horrour' they may well have had Wastwater in mind. But most of the central valleys of Lakeland must have exhibited the starkness of Wasdale before the extensive tree-planting of the nineteenth century, if one is to believe such travellers as Defoe who visited Lakeland early in the eighteenth century: 'Nor were these hills high and formidable only, but they had a kind of unhospitable terror in them . . . but all barren and wild, of no use or advantage either to man or beast.' Even the early writings of the Romantic poets reflected a mixture of fear and fascination for the craggy central fells, as exemplified by Coleridge who climbed Scafell in 1802 and described its rocky face as '. . . the frightfullest Cove that might ever be seen . . . just by it and joining together, rise two huge Pillars of bare lead-coloured Stone . . . their height and depth is terrible'. A mere twenty years afterwards these same poets were looking in more detail at the character of the rocks which were now beginning to inspire them. Keats, for example, after visiting Stockgill Force near Ambleside, discovered that 'what astonishes me more than anything is the tone, the colouring, the slate, the moss, the rockweed . . . I shall learn poetry here'.

The earliest settlers in the Lake District were not as concerned about the colours of the volcanic rocks as they were about their hardness and the ease with which they could be fashioned into tools. Prehistoric man first settled in Cumbria on the coast, where he lived by fishing and hunting. By New Stone Age times, however, these primitive tribes had penetrated inland and had discovered that certain of the volcanic rocks could be worked into rough artifacts; before long stone-axe factories were operating on the slopes of Great Gable, Scafell and the Langdale Pikes. So successful were their products that they were traded all over England and Scotland. Nonetheless, the inhospitable mountainland attracted

58

A landscape in
the Lake
District, where
glaciers have
scoured the
rugged
Borrowdale
volcanic rocks.

'Wasdale,
whose silvery
Wastwater is
surrounded by
beetling crags.'

Langdale,
where Stone
Age man used
Lakeland's hard
volcanic rock to
fashion his early
artifacts.

virtually no permanent settlement even in the succeeding Bronze Age, notwithstanding the two well-preserved stone circles of Castlerigg (near Keswick) and
Swinside (near Millom) both of which are peripheral to the central fells. It is true
that some Iron Age hill-forts, such as that on Castle Crag in Borrowdale, were
erected in the volcanic fells but their diminutive scale reflected both the broken
nature of the terrain and the smallness of the social unit when compared with the
great hill-forts of the less dissected Wessex chalklands with their large tribal
groupings. The stony Lakeland fells remained desolate even during the Roman
occupation for they were part of their military zone; the Roman road that strikes
westwards from Ambleside to the coastal fort at Ravenglass is still a lonely one.
Today, the grey ruins of Hard Knott fort, high in the windy western fells, conjure
up visions of Roman legionaries far from their sunny homeland.

Not until the Viking invasions of the Dark Ages was the Lake District to
become permanently settled and it was the Norsemen who gave to Lakeland
most of its distinctive names, names which tell us something of the difficult
environment that they encountered and of the impact they had upon it. The term
'thwaite', for example, means 'clearing in a forest' and as a suffix it is common-
place throughout Cumbria. The village of Stonethwaite in Borrowdale conveys
something of the difficulties involved in settling this boulder-choked valley, for
the Vikings were sufficiently astute to place their hamlets on poorer ground in
order to reserve the fertile soils for their hay meadows and arable plots. They

Brownside Farm, Dunnerdale, Lake District. A typical longhouse with unpainted byre and barn.

also had an eye for water supply and safety, as exemplified by the village of Troutbeck ('trout stream') strung out along a valleyside springline and above the winter flood-level of the stream.

Today, many of the Lakeland farmhouses and cottages still occupy these early sites, possibly built with some of the same stones first cleared from their primitive fields by the Norsemen. The oldest of the traditional buildings is the longhouse, in which house, byre and barn were constructed under one roof. The rubble stone was generally uncoursed, with the lumpy volcanics laid at random but with great dexterity. As in the granitelands, the boulders and cobbles were retrieved from the stream beds and from the open moors and in many of the barns and humbler dwellings the stones were laid without mortar. Only the more vulnerable chimney stacks were roughly mortared but in the older buildings these are distinctively circular to avoid having to make corners from difficult stones. Where slaty rocks were at hand, however, they were often used for door jambs and lintels. The gentler fells have provided slates for local roofing and wall-cladding throughout the region, especially from the Coniston, Elterwater and Kirkstone quarries.

Present-day inhabitants of the Lakeland farmsteads have maintained the age-old custom of cement-rendering and whitewashing the house walls but leaving the rough stonework of the outbuildings uncovered and unpainted. Although tourists find the white cottages picturesque it is the untreated masonry of the ancillary buildings that conveys the true character of volcanic Lakeland. 'They may rather be said to have grown than to be erected – to have risen, by an instinct of their own, out of the native rock – so little is there in them of formality, such is their wildness and beauty'. So wrote Wordsworth in his *Guide to the Lakes*, and it is his voice which must be given the final word in summarizing the way in which characteristic Lakeland architecture fits harmoniously into this unique environment. 'The stones are rough and uneven in their surface, so that both the coverings and the sides of the houses have furnished places of rest for the seeds of lichens, mosses, ferns and flowers. Hence buildings, which in their very form call to mind the processes of Nature, do thus, clothed in part with a vegetable garb, appear to be received into the bosom of the living principle of things, as it acts and exists among the woods and fields.'

## The Great Whin Sill

The Lake District's eastern neighbour, the Pennine range, is built from thick, relatively unfolded, layers of limestone, shale and sandstone. Although these sedimentary rocks have not been greatly deformed by folding, the whole massif of the Northern Pennines has been lifted up in the west and tilted down in the east. The same crustal upheaval which raised this extensive plateau into the mountainous bulk of Cross Fell, the Pennines' highest peak, also witnessed a limited movement of magma not far below the surface, as volcanic lavas strove to burst forth. Though the landscape was never flooded by lava flows, extensive sheets of dolerite were squeezed between the beds of limestones in the form of a gigantic sill. Where erosion has subsequently exposed this dark volcanic rock at the surface, it has helped to create a remarkable phenomenon: the intrusive black step in the gentle white staircase of Pennine limestones is known as the Great Whin Sill.

'Whin' and 'whinstone' are quarrymen's terms for any dark-coloured rock, such as dolerite or basalt. The stone is so much harder than its neighbouring limestones and sandstones that wherever this dolerite intrusion is exposed it forms a vertical cliff, often over 50 feet in height. The Whin Sill outcrop stretches for more than 100 miles in a broken, curving line from the deserted moorlands of Upper Teesdale and Cross Fell, eastwards across the Tyne valley to the beautiful coast of Northumberland (see Figure 7). Almost everywhere its distinctive character has produced striking scenery, ranging from the spectacular waterfalls of Teesdale, through the historic tracts of Hadrian's Wall to the picturesque sea cliffs of Bamburgh Castle and the Farne Islands.

High Force,
North
Pennines.

Figure 7 The
Great Whin
Sill.

High on the windswept plateaux of the Pennines, just below the summit of
Cross Fell (2,930 feet), the infant river Tees begins its long journey to the North
Sea. After flowing gently through lonely grasslands and the controversial Cow
Green Reservoir (where a unique mountain flora was inundated by an insensi-
tive regional water authority) the river first encounters the outcrop of the Whin
Sill at Cauldron Snout. The gentle meandering stream in its broad limestone
valley is suddenly transformed into a torrent as it plunges over the layer of dark
volcanic rock in a series of cascades. A few miles downstream the Tees crosses
the Whin Sill once more but this time it plunges some 70 feet in one spectacular
leap at High Force, one of England's largest and most dramatic waterfalls. The
dark-coloured, well jointed rock band at the lip of the fall has withstood the
constant attack of the peat-stained waters much more successfully than have the
bedded limestones beneath. Where the base of the waterfall has quarried away at
these lower, less resistant beds it has undermined the dolerite band and left it
precariously overhanging. Periodically, masses of the black whinstone crash
down into the bouldery gorge causing the waterfall to retreat gradually upstream.

The Roman Wall, west of Housesteads, Northumberland, with tumbled whinstone in the foreground.

Farther north, near to Hexham, the line of the Whin Sill crosses the valley of the River Tyne and for many miles on either side of this important trans-Pennine routeway the north-facing whinstone cliff creates a natural bastion facing the Scottish border. Not surprisingly, in a region where border warfare prevailed for centuries, this critical tract of countryside has retained many examples of military architecture, ranging from the remarkable bulwark of the Roman Wall, to the castles and fortified houses of Northumberland. Hadrian's Wall, commenced in AD 123, marked the existing northern frontier of Roman Britain and its line of ruined defences can be traced for 73 miles between the Solway Firth and the mouth of the River Tyne. Because stone was lacking in the Solway marshes the western limits of the wall were originally built from turf but once it had climbed to the Pennine heights the Romans could choose from a variety of rocks to raise their 15-foot wall and construct their mile castles and forts to house

The carefully tooled masonry of the Roman Wall at Corbridge, Co. Durham.

their troops. Most commonly they selected the easily quarried buff-coloured sandstones or white limestones but occasionally dark whinstones were incorporated in the well coursed masonry. They soon appreciated the defensive opportunities offered by the Whin Sill and for lengthy stretches the wall follows the dolerite cliff as it marches across the fells. Today's landscape is one of undulating hills and deeply cut valleys, a peaceful place of open skies, lonely, sheep-grazed moorlands and rough pastures, echoing to the song of the skylark and the bubbling call of the curlew. Nevertheless, the implacable line of the wall is a constant reminder of the strife which long characterized the history of this Border country. Its presence seems to amplify the intrusive nature of the volcanic rock and illustrates how a combination of human and natural forces produced a barrier against the northern intruders.

For centuries after the departure of the Romans the wall's stonework has been plundered for use in farmstead and field wall. Throughout the troubled years of the Middle Ages its masonry was incorporated into the solitary peel towers which served as fortified farms. They were built with the living quarters on the two upper storeys, above a ground-floor room devoted merely to storage. The three-storey peel tower, built from well coursed stone, can be compared with the two-storey defensible farmhouse, or bastel-house, in which livestock occupied the vulnerable ground floor. In both cases entry to the upper floors was effected by a ladder, though external stone staircases were ultimately added after political stability was restored in the eighteenth century and agriculture given a chance to prosper.

Where the Whin Sill reaches the North Sea coast it forms a line of columnar sea cliffs between Alnwick and Bamburgh before it disintegrates into a scatter of bird-haunted reefs and stacks known as the Farne Islands. The much altered but still imposing Bamburgh Castle on its whinstone crag above the dune-fringed beach is perhaps the best known of the coastal castles, although the ruined fortress of Dunstanburgh is the most atmospheric as it perches forlornly on its dolerite promontory jutting into the grey North Sea. Pevsner found it '. . . one of the most moving sights of Northumberland' and its ruined silhouette has long been a favourite subject for artists, most notable of whom were Turner and Thomas Miles Richardson. The vertically jointed structure of the dolerite and the exposed flat-topped surface of the sill have combined to create a multitude of nesting sites for the thousands of guillemots, kittiwakes, shags and puffins, making the Farne Isles one of Britain's most valuable nature reserves. In addition to their function as a haven for seabirds, the whinstone ledges have also provided a breeding ground for some 7,000 grey seals, the sole example on Britain's north-eastern coast.

## The Hebrides

Nowhere is volcanic stone more assertive than in the islands of Mull and Skye, off the western coast of Scotland. The volcanic rocks of Snowdonia and Lakeland were formed 500 million years ago and the Whin Sill intruded 300 million years ago, but the Hebridean volcanic episode culminated only 50 million years ago, so that its lava flows, ash bands, sills and dikes are relatively 'fresh' by geological standards. Furthermore, although the volcanic rocks have been heavily scoured by ice sheets, they are so widespread and all-enveloping that they dominate the scenery not only of these two remarkable islands but also of their smaller neighbours.

In Gaelic the name Mull signifies a 'high, wide tableland' and so far as the northern part of the island is concerned this is an apt description, for here successive lavas have been piled upon each other until an enormous basaltic shield, some 6,000 feet thick, has been forged. Rivers and glaciers have subsequently carved the lava plateau into a series of flat-topped hills and intervening valleys, along whose sides structural benches, created by the horizontally

A landscape of lava, Isle of Skye, Scotland.

disposed lava flows, can be traced for miles to create what is known as a 'trap' landscape (from the Swedish 'trappa' meaning 'step'). The stepped character of the basaltic slopes is also characteristic of Skye and other basaltic islands. Each of the 'treads', having weathered into a gently sloping terrace on the more friable upper surface of each lava flow, is now covered with peat and a heathery vegetation. Separating the treads of this volcanic staircase are the steeper, and sometimes precipitous, 'risers' which represent the more resistant lower layers of each flow. Dr Samuel Johnson spoke of Mull's northern plateaux as 'a gloom of desolation' and suggested that they should be planted with trees to give them a 'more cheerful face'. Some two centuries later the Forestry Commission has duly obliged, much to the dismay of farmers and conservationists who contend that the mineral-rich basaltic soils, where adequately drained and managed, could produce more worthy crops than conifer trees. Sir Frank Fraser Darling, the

noted naturalist, agrees that the Hebridean basaltic soils are capable of producing 'the most fertile land in the Highlands and some equal to the best in the
kingdom'. Referring to the neighbouring, almost uninhabited volcanic islands of
Ulva and Gometra, now choked with bracken and hazel scrub, Darling com-
plains of the poor husbandry: 'the Highland paradox is apparent again, in that
here are green acres gone derelict while there is still congestion on the poverty-
stricken Archaean gneiss of Lewis'. The explanation is to be found in the
harshness of the crofting life in these isolated communities of the Celtic fringe.
Apart from the constant wind and heavy rainfall of these Atlantic coastlands, ice
sheets have stripped the nutritious soils from all but a few favoured sites. Above
all, the ubiquitous basalt has given a stony countenance to every hillslope and
tableland except where blanket peat has softened them. Little wonder that many
of the stone cottages are now abandoned, and their former inhabitants settled
overseas in Canada, Australia and New Zealand.

A mere handful of whitewashed cottages, with their crofting families, has
survived in the so-called 'Wilderness' of the Ardmeanach peninsula of south-
west Mull. Their tiny hay plots provide splashes of yellow in the drab colouring of
the sheep-dotted moorlands, where dwellings stand windswept and isolated by
the rocky shore or seek shelter beneath the steep basaltic cliffs of this lonely
terrain. Few of the tourists who pass this way, probably en route to the fabled isle
of Iona, will be aware of another remarkable survival on this rugged coast. A
fossil tree, silicified into a 40-foot column of glistening quartz crystals and
surrounded by a sheath of black charred wood, was discovered in 1819 by a
certain Mr MacCulloch after whom it has been named. Its roots are bedded in a
band of red volcanic ash and its trunk is almost buried by the enveloping basalt.
This 50-million-year-old tree, seared into stony immobility by a sudden flood of
lava, somehow encapsulates the violence which once shattered the tranquillity of
these Hebridean environments, just as the recent eruption of Mount St Helen's
in North America destroyed the majestic forests which flourished on its formally
tranquil slopes. Although the Hebridean scene is now calm, one cannot ignore
the tormented volcanic landforms which impose themselves at every turn of
Mull's twisting roads. Yet the most dramatic of all lies offshore, on the tiny isle of
Staffa.

Mendelssohn's well known music, portraying the varying moods of Fingal's
Cave, has made the cathedral-like sea cave one of the most famous of British
landmarks. Its basalt portals have been painted by such eminent artists as Turner
and Copley Fielding, eulogized by Scott, Keats and Wordsworth, and subsequently photographed by countless visitors, in the two hundred years after its
'discovery' by Sir Joseph Banks. Staffa's aesthetic appeal is based entirely on the

Fingal's Cave, Staffa, Scotland.

majesty and symmetry of these columns, for the structural differences within a single lava flow, reflecting the contrasting rates of cooling, are nowhere better displayed. The slaggy crust cooled rapidly from above to produce an almost structureless capping, which now forms the roof of Fingal's Cave; the middle part of the flow also cooled from above, but its slower and more regular rate of shrinkage led to the formation of tall, narrow polygonal columns (like the piers of a cathedral nave); the lowest part of the flow cooled from below, with its very slow contraction producing the massive, regularly spaced columns of the 'causeway' which leads into the cave. The architectural character of basalt rock formations is such that in 1694 the scientists of the Royal Society thought it necessary to devote time to confirming that the columns of Ulster's Giant's Causeway were not man-made, because of the lack of mortar in the joints!

Few place-names in Britain conjure up such romantic visions as that of the 'Misty Isle' of Skye. It is an island of superlatives: the largest island of the Hebrides; with the greatest expanse of basalt plateaux in Britain; it boasts the most spectacular mountain group of the British Isles. Skye is a land of bold peninsulas where sea lochs bite deeply into the interior, and there are those who claim that its name is derived from '*sgiath*', the Gaelic word for 'wing' – hence its soubriquet the Winged Isle. Although the jagged ridge of the Black Cuillins is the most dramatic landform in the island its rock is a dark plutonic gabbro which, like granite, cooled into a massive bulky dome deep below the crust. The lava terrain, where volcanic basalts poured out across the land surface, is found in the northern half of the island. Here the trap landscapes of Mull are repeated on a grander scale, for the ocean has carved the headlands into spectacular sea cliffs. But not all the great precipices of the northern peninsulas have been created by marine agencies – the east-facing escarpments of Trotternish, for example, were formed in a very different way.

Between the attractive town of Portree and the isolated northern village of Staffin a 20-mile line of fearsome precipices stands back some distance from the shore to form one of the most remarkable of the British landscapes of stone. The countless layers of basaltic lava which, in the interior, build such flat-topped peaks as MacLeod's Tables, have, in eastern Trotternish, been tilted upwards to form the lofty rock faces of the Storr and the Quiraing. Moreover, when the lavas spewed forth they buried not a floor of hard primeval rock but substantial layers of bedded limestones and clays similar in age to those that build the gentle Cotswolds. Because these weaker sedimentary rocks have been unable to bear the weight of the stupendous lava pile, the upturned plateau edge has collapsed into a confused topography of landslips, the most extensive landslide terrain in Britain. At their base, the 1,000-foot cliffs of the Storr and the Quiraing have disintegrated into an unstable clutter of jagged buttresses, pinnacles, leaning towers, house-sized boulders and blocky screes. Though visitors may be fascinated by the leaning spire of the Old Man of Storr or the black Needle Rock in the Quiraing, they may also feel as if they had strayed into a volcanic crater as they survey the gigantic amphitheatre of black rock where tier upon tier of lava climbs to the jutting skyline. Nowhere in Britain are volcanic landscapes so majestically overwhelming. If Fingal's Cave needed a Mendelssohn to capture its *genius loci* then the Trotternish scarps merit a Wagner.

It is hardly surprising that such barren and hostile terrains deterred any prehistoric settlement, which meant that the early Celtic monks, dispersing from Iona, were among Skye's first permanent inhabitants. Paradoxically, the daunting peninsula of Trotternish was chosen as the site of the first Celtic church – on a tiny island in Loch Chaluim-chille (Malcolm's cell). Today the loch has been drained and the historic place is little more than a stony mound surrounded by fragrant hay meadows overlooking the sea. But archaeologists have shown that within the drystone perimeter wall and its associated earthen bank this primitive monastery still retains the foundations of two small chapels and a cluster of tiny ruined 'beehive huts'. A similar collection can be seen at Annait, not far from the dark stone castle of Dunvegan on its basalt sea cliff. The ingenious corbelled design of these tiny drystone structures illustrates the dexterity of the early islanders, for the lumpy volcanic stone is not easy to manipulate. Yet, below the frowning basaltic cliffs the low coastal plateaux of Trotternish are underlain by limestones and clays, the nutritious soils of which were to support a substantial population in northern Skye.

The Gaelic people of the Hebrides were never urban dwellers but preferred to

The jagged
gabbro ridge of
the Black
Cuillins, Skye,
Scotland.

The basaltic
buttresses and
pinnacles of the
Storr, Skye,
Scotland.

Evolution of the
Hebridean
Cottage.

live in small scattered tribal groupings known as 'clachans'. So closely knit were
these Celtic communities on the remote islands that their crofting way of life has
survived to the present century. Their dwellings have evolved from the simple
monastic beehive huts and, like those on the granitic coast of Cornwall, have
been designed to withstand the fierce Atlantic storms. The detailed character of
the cottages was determined by the availability of local materials for their
isolation was such that no building stone could be imported, at least not until the
advent of the Victorian lairds many centuries later. Equally, the treelessness of
these western isles has meant that the roof timbers were among the crofter's most
valuable assets, to the extent that they were used in house after house during
periodic rebuilding. Since slates and tiles were unavailable, the roofs were first
covered with branches upon which overlapping layers of turf were placed,
pegged with heather stems and tied down with ropes of plaited straw known as
'sugan'. Only later were these primitive structures thatched and secured by hemp
ropes weighted with pebbles. The thatch was usually of barley or oat straw, taken
from the tiny patchwork of arable plots, although the most primitive were roofed
with a heather thatch. There are few survivals of these earliest 'black houses', as

they were termed, which were constructed as little more than crude enclosures, squat and streamlined, with rounded corners to give minimum resistance to the wind. Their primitive appearance was such that eighteenth-century travellers dismissed them as dark, smoking dunghills. On his famous tour of the Hebrides in 1773 Dr Johnson describes how 'the traveller wanders through a naked desert . . . and now and then finds heaps of loose stones and turf in a cavity between the rocks, where a being is condemned to shelter itself from the wind and the rain.' Some thirty years later, Coleridge, Dorothy and William Wordsworth painted a less gloomy picture and at the same time explained how the chimneyless 'black house' earned its name. These intrepid tourists sat away from the livestock, around an open fire burning in the centre of the mud floor '. . . observing the beauty of the beams and rafters gleaming between the clouds of smoke. They had been crusted over (until) they were as glossy as black rocks on a sunny day cased in ice.' Ultimately the earliest type of black house, with its double walls of undressed stone, evolved into a more substantial dwelling in which the byre was subdivided from the living quarters and roughly squared volcanic stones were utilized in lintels and door jambs. Finally, when squared corners, gable ends and chimneys were introduced, in the so-called Dalriadic house, the Skye cottages became comfortable dwellings whose dazzling white façades now 'look out like kindly faces from the brown hillsides'.

# LIMESTONE LANDSCAPES

There is no more striking contrast than that which exists between the hard, dark volcanic landscapes of western Scotland and the soft, pale-coloured limestone tracts of England. With the former, the dour terrains of rough-textured, irregular volcanic rocks almost everywhere give a rugged, broken line to the landscape. With the latter, the scenery is characterized not only by its flat or gently curving skylines and its lengthy scarps but also by its sweeter calcareous soils which support verdant pastures on the hills and boundless prairies of golden grain at lower elevations. But limestones also have certain properties which distinguish them from other rocks. First and foremost, because of their bright-toned minerals, limestones reflect the light. Their grains of calcite sparkle in the sun and rain, while their various hues evoke either the rich qualities of cream and honey or the cool grey tones of a dove's plumage. Even on overcast days the bare limestone 'scars' of the Pennines or the chalk cliffs of the Channel coast bring a certain luminosity to the scene. Its second characteristic, an inability to retain water, means that limestone soils are almost everywhere free-draining; water shortages can threaten crops to such an extent that, traditionally, the limestone grasslands of England have been devoted to sheep rearing (though modern farming practices have brought sweeping changes on both wolds and downs). For similar reasons settlements have been forced either to cluster where springs break out at the scarp foot, or to be strung out along the deeper valley floors where streams flow perennially; the higher valleys are often streamless, as the chalklands most clearly illustrate. Rarely are towns and villages sited on the hill tops of limestone country; the few exceptions, like Monyash or Stow-on-the Wold, stand as isolated as oases amidst the virtually waterless countryside. The third feature of limestone, and the one that has had the most important effect on both the physical and the cultural scene, is the presence of well developed joint patterns within the bedded stone. In combination with limestone's natural porosity, it is the vertical jointing which helps to lead rainwater quickly underground and produces such a paucity of surface streams. The few major rivers which manage to cross Britain's limestone belts are fed, therefore, by a network of underground streams flowing through subterranean passages, pipes and caverns, all channelled along the bedding planes and joint mosaics of the

(*Above*) Mountain Limestone cliffs in the Cheddar Gorge, Mendips, Somerset.

(*Above right*) The chalk cliffs of the Seven Sisters, Sussex.

stratified rocks. Without such constant underground replenishment it would be impossible for the Dove and Derwent, for example, to traverse the parched terrains of the White Peak, or for the 'bourne' hamlets to survive on the bleached chalk downs. But the vertical jointing has also affected limestone scenery in other ways: the perpendicular structures help the relatively soft chalk to stand precipitously at Beachy Head and the Mendip limestones to form the towering walls of Cheddar Gorge. The same well defined lines of weakness have aided rivers in their never-ending quest to incise their valleys deeply into the calcareous rocks: the Avon gorge at Bristol, the Wye at Chepstow, the Witham at Lincoln, the Thames at Goring, the Wey at Guildford are but a handful of examples. It is significant, however, that each of these limestone outcrops has few if any other surface streams on the flanks of its narrow water gaps.

Quarrymen in the limestone belts from the most primitive times have appreciated how the joint patterns have also facilitated their own efforts to win blocks of

One of Britain's oldest limestones, the Wenlock Limestone, was used in Much Wenlock Priory.

building stone, even if their operations pale into insignificance when compared with the broad-scale sculpturing of nature. Good quality limestone has been sought by British stonemasons since Roman times and it is this, more than any other stone, that has played a paramount role in the architecture of town and country alike. Not only can it be easily quarried and sawn for use in church, mansion or palace, but the surplus rubble and 'waste' can also be utilized in cottage, barn and fieldwall. It is one of the strongest of the sedimentary rocks and, properly quarried and laid, limestone is fairly weather-resistant building material. Above all, a good deal of it is amenable to intricate carving which means that many English cathedrals exhibit a spectacular display of pinnacles, tracery, arches and statuary, all carefully chiselled from numerous varieties of this pale but venerable stone. Such differences of texture and colour in the building stones are often echoed by the various landscapes from which they have been quarried. The dove-grey shelly stone, which gives such staunch solidity to Shropshire's Wilderhope Manor and Much Wenlock Priory, for example, extends its ruler-straight outcrop from Ironbridge to Craven Arms as the beautiful Wenlock Edge. The rugged, steadfast masonry of Beaumaris Castle matches the jutting but heavily quarried nearby headland of Anglesey's Penmon Point. Each limestone landscape has its own distinctive landforms and buildings: those of the Pennines, the Cotswolds and the chalklands are particularly noteworthy.

# The Pennines

There is a marked distinction between the younger and generally softer limestones which build the ridges, downs and wolds of the English lowlands and those older, more massive, thickly bedded and compact limestones from which the bolder landscapes of upland Britain have been carved. The contrasts are such that the older, harder limestones have been collectively referred to as Mountain Limestone, most of which was laid down in tropical seas just before the great Coal Measure forests flourished in the coastal swamps and river deltas. The gradual transition from clear warm ocean through a period of sandy and muddy deposition to one in which thick rain forests spread across an emerging land surface has left an indelible record in the thick sedimentary accumulation which now forms the Pennines, popularly known as the 'backbone of England'. At the base, the corals and shells have been cemented into massive layers of Mountain Limestone; the sandy deposits now form the tough Millstone Grit and the muddy sediments have been compressed into shales. Overlying all are the Coal Measures in which the decayed forest vegetation has been converted into seams of coal which themselves are interleaved with shales, grits and thin limestones, indicative of a periodically flooded coastline. Some 300 million years ago the entire sedimentary succession was heaved into a gigantic upfold, since when erosion has worn away the uppermost rocks on the crest. Rivers have cut down so deeply that in the central sectors of the Pennines the pale Mountain Limestone has been uncovered and is now overlooked by flanking scarps of dark Millstone Grit. The Peak District of the southern Pennines mirrors the differences in colour between the limestone and the gritstone for it has been subdivided into two regions – the White Peak and the Dark Peak.

The White Peak of Derbyshire, with its bright green pastureland, grey-white cliffs and mosaic of gleaming drystone walls, stands out like a scintillating jewel amidst the sombre moorlands of the surrounding gritstone. Not far beyond the moors lie the urbanized coalfields of the industrial North. Yet the solitude of the White Peak has survived and its tiny stone-built settlements stand virtually unaltered since they were painstakingly erected several centuries ago. Although these isolated villages harmonize with the natural scenery, they are so sporadic that it is the fragmented pastures, the deeply incised gorges and the bare-faced scarps that dominate the scenery. Because the Pennine plateaux are higher than most of the chalklands or the Cotswolds their vegetation is less luxuriant and the trees more wind blown. This is no place for cornfields or even root crops, the climate is too spartan and the soils too thin for anything but sheep-grazed turf.

The entrance to
Dovedale,
Derbyshire.

Nonetheless, the sweeter soils support a profusion of colourful wildflowers that
would not be found on the sour, peaty moorlands of the neighbouring grits,
indeed the White Peak supports one of the richest floras in central Britain. At
various locations one may find colonies of rich blue Jacob's Ladder, pale-blue
Scabious, White Aramis and Lily-of-the-Valley, the delicate Mountain Pansy
and the Glaucous Rose, to say nothing of the reds and pinks of Campion and
Bloody Cranesbill or the yellow Globe Flowers and the Giant Bellflowers of the
damp ashwoods. Woodlands flourish mainly in the sheltered gorges whose
defiles are choked with a tangle of ash, elder, hawthorn and hazel through which
the rivers swiftly glide. The bare, grey rock faces of Dove Dale, Monsal Dale and
Miller's Dale, for example, where isolated pinnacles overlook the murmuring
waters, are also dotted with white hornbeams or dark-leafed yews. In the villages
themselves occasional sycamores bring a softer texture to the hard lines of the
roughly dressed masonry.

In contrast to the mellow Cotswold oolites, the Mountain Limestone rarely provides a good freestone and its quarried beds have been unable to produce a supply of cleanly squared blocks, which means that its lumpy stones are more fitted for rubble walling than for smooth-faced ashlar. Furthermore, because it is considerably tougher than oolite, the Pennine limestone does not lend itself so easily to detailed carving and is therefore rarely used for village churches. Nevertheless, the sturdy White Peak farms and cottages bear a freshly-scrubbed look and their radiance in full sunlight simply amplifies their close affinity with the gleaming rocks and glinting fieldwalls of this luminous landscape of stone. Near the regional boundaries, however, the more easily worked and darker-toned gritstone has been extensively used for corner stones, jambs and lintels, with sandstone flags utilized as a roofing material in the traditional limestone dwellings. This is particularly true in the humble industrial village of Bonsall, whose seventeenth-century rubble-walled, three-storeyed houses cluster pic-turesquely around a sandstone-stepped cross. The lead mining which caused Bonsall's former prosperity has now ceased everywhere in the Peak District, though its surface scars and spoil heaps can still be discerned, often marked by long lines of trees planted to help disguise the spoliation. In villages such as Hartington, above Beresford Dale, only the older houses are built of limestone but this gives a cohesive dignity to those that crowd irregularly around its green. Limestone is used exclusively in the centre of the White Peak to build such attractive villages as Thorpe, Tissington and Parwich, though to recapture something of the atmosphere of former times one has to travel to hoary old hamlets like Flagg or Alstonefield which, set high on the bare uplands amidst a chequerboard of fieldwalls, have a stony echo to their names. Almost all of the Peak's symmetrical drystone walls were geometrically laid out during the eighteenth-century enclosure acts. Near Castleton, however, a group of narrow, linear fields are bounded by walls that have a flattened S-shaped plan, indicating that they were created merely by enclosing the ancient strip-fields which had survived since Domesday when they were part of the common-field system of feudal land tenure. The sinuously curving walls which enclose the water meadows of the river valleys are, by contrast, a measure of the terrain rather than of farming practices, while the linear villages of Winster and Youlgreave, for example, grew along a valley or springline simply in their quest for water.

One of the best known of these river valleys is that of the Dove (Fig. 8(a)), about which guide books regularly eulogize: 'There is no river in England to compare with the Dove (yet) in spite of its peerless beauty it is the most modest of rivers, hiding itself in a deeply cleft passage of which the whereabouts cannot

Litton, in the White Peak of Derbyshire, with its mosaic of limestone walls.

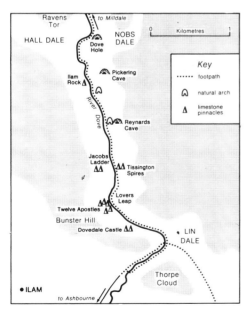

Figure 8(a) Dovedale, Derbyshire.

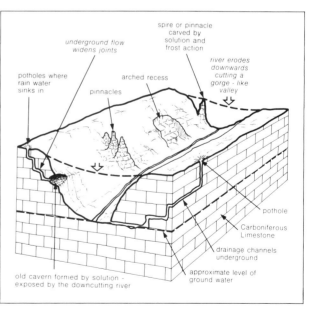

Figure 8(b) Mountain Limestone landforms, Derbyshire.

Drystone
walling in the
limestone
Pennines.

Rainwater
etching of
Mountain
Limestone
creates patterns
to match the
gnarled ash
trees.

readily be perceived'. The 'deeply cleft passage' is, of course, Dove Dale where Izaak Walton described the beauties of the gorge '. . . while seeking to beguile the wary trout . . . or scaling the steep ascent to Reynard's Cave, or standing with wondering admiration before such mighty monoliths as that called Ilam Rock, or the Pickering Tor'. There are many other spires and pinnacles in the dale . . . 'frowning in craggy grandeur and shaggy with the dark foliage that grows out of chinks and clings to the asperities of the rocks'. Figure 8(b) illustrates how some of these remarkable landforms were created partly by frost and partly by swollen waters melting from an ice sheet during the Ice Age. But none of these spectacular natural features of spire, arch and cave could have been sculptured without the existing network of underground drainage that had been slowly carved out of the massively jointed Mountain Limestone. In fact, the downcutting river Dove has merely discovered the subterranean maze of passage ways and exposed it to the sky. Nonetheless, to examine the finest examples of Britain's underground stream patterns one has to travel farther north to the Craven Uplands and the beautiful limestone landscapes of the Yorkshire Dales.

The scenery of the northern Pennines is dominated by Mountain Limestone and because its outcrop is considerably more extensive than in the Peak District the crags, scars, sinkholes, caves and gorges are more widespread. Where the massively bedded Great Scar Limestone occurs around Ingleborough and Malham, the landscapes achieve a monumental scale of stony grandeur. Here the beds of greyish-white, shelly limestone are overlain by layers of muddy shales, thin limestones and sandstones (known collectively as the Yoredale Series of rocks) that build the impervious capping from which former rivers and glaciers have carved out the isolated peaks of Ingleborough, Pen-y-ghent and Whernside. The darker rocky summits of these Pennine giants rise like truncated pyramids above the gleaming limestone plateaux. For years it was thought that Ingleborough was Britain's highest mountain.

Whereas the narrow defile of Dove Dale escaped the bulldozing effects of a valley glacier, each of the Yorkshire Dales had its own personal glacier which opened out the valley profile and created the characteristically steep sides and wide floors seen today. The bare white cliffs ('scars') which flank the present dales were chiselled out and scraped by moving ice: Kilnsey Crag, Raven Scar and Twistleton Scars are among the most remarkable. Above them, on the higher valley slopes, rise veritable staircases of grassy 'treads' and gleaming 'risers' from which over-riding ice sheets have scoured much of the soil and exposed naked limestone pavements. The best known of these stands at the brink of Malham Cove's beetling cliffs, its scrubbed surface criss-crossed with deep fissures ('grikes') where rainwater has widened the joints and left the

'The darker
rocky summits
of these
Pennine
giants rise like
truncated
pyramids above
the gleaming
limestone
plateaux.'

Kilnsey Crag,
Wharfedale,
North
Pennines.

The clints and
grikes of
Malham Cove's
limestone
pavements.

intervening calcareous rock as a fretwork of pedestals ('clints'). Although most
limestone pavements seem devoid of vegetation, a rich hidden flora thrives in
these fissures, protected from the vagaries of wind, sheep and man.

Once the surface water has percolated underground it continues to attack the
limestone along every line of weakness, thereby creating, over thousands of
years, a maze of subterranean caverns. All the streams which descend steeply
from the shaly summits of the great peaks soon cross onto the plinth of porous
limestone where they rapidly disappear into shafts known as potholes. The most
renowned is Gaping Gill, on the flanks of Ingleborough, where the Fell Beck
plunges vertically for some 350 feet. Finally, all these underground streams
eventually reappear in the valleys far below, either as a quiet 'rising' in a dark
placid pool or as a chattering brook from a cave mouth, like that at Clapham
Beck Head or at Malham Cove. The enormous amphitheatre of Malham Cove,
with its 300 feet of bare, precipitous limestone, requires a special explanation,
for vertical cliffs of this magnitude are rare in Britain away from the glacially
eroded mountainlands. Above the cove a dry valley can be traced back to lonely
Malham Tarn, but its waterless channel must once have been filled by a surface
stream. During the Arctic conditions of the Ice Age, however, all underground
water would have been frozen and the frigid cave network sealed by ice. Surface
waters from the melting ice sheets of the Craven uplands would then have had to

Malham Cove,
Yorkshire: the
site of an Ice
Age waterfall.

carve out an alternative route to the edge of the cove's faulted escarpment over
which they must have plunged as a mighty waterfall. Only in post-glacial times,
when the climate had ameliorated, could the streams return to their underground courses. At neighbouring Gordale Scar the limestone roof of a similar
subterranean cave system has collapsed to create an awesome chasm of thunder-
ing waterfalls and overhanging walls of rock. The eighteenth-century poet
Thomas Gray spoke of the 'savage aspect of the place' and described how he
shuddered for a quarter of an hour beneath 'the dreadful canopy' of the
overhang. Such sublime scenery brought some of Britain's most famous artists to

Gordale Scar, Yorkshire: a collapsed cavern (from a 19th century painting by W.J. Müller, R.A.).

A crystal clear stream in the Yorkshire Dales.

Malham and Turner, Muller and James Ward were able to portray the overwhelming majesty of Gordale Scar in their own inimitable styles.

The crystal clear waters of Malham Beck, reappearing surreptitiously at the foot of the cove, finally abandon the scarified and desiccated upland scenery as they wend down towards the verdant tracts of Airedale. Here, Malham, with its packhorse bridge and limestone cottages, is typical of the attractive villages which make such a major contribution to the Yorkshire Dales' scenery. Unlike the White Peak of Derbyshire the Dales have no settlements on their uplands; the villages crowd into the sheltered valleys, generally along the springlines. Few would dispute that the landscapes of the Dales gain an added distinction from their buildings, which are constructed mainly from roughly coursed and crudely dressed rubble. But whether it be a ruined abbey, stately manor, robust farm or

Isolated farms and barns in the Dales.

humble cottage, all are dignified by the cool gleaming limestone. The pale-grey masonry and lichen-crusted roofs of local flagstone create a simplicity of tone but a nobility of proportion which together evoke a sense of unified rectitude in this spartan landscape. Not surprisingly, some of Britain's greatest watercolour artists, including Turner, Girtin and Cotman were inspired by the pastel tones of the Yorkshire Dales.

Nowhere is the harmony between man and nature more clearly illustrated than in Swaledale and Wensleydale, where the settlement pattern of clustered villages, isolated farms and lonely field barns ('laithes') is a legacy of the Norse colonization of the Dark Ages. The dove-grey limestone has been utilized in virtually all the older buildings, with not a single red brick or roof tile to be seen. In their excellent little book *Houses in the Landscape* the Penoyres have captured the essence of the scene: 'The Arcadian effect of the landscape is heightened by the form of the buildings, which have low-pitched roofs and a direct simple walling technique of projecting "through" stones used almost

universally in the outbuildings and barns and sometimes in the houses too . . .
The effect is reminiscent of Alpine or even Italian farm buildings.' Yet the
austere beauty of the Dales reflects more of ice-scoured Scandinavia than the
sunnier climes of southern Europe, and this is borne out by their Viking place
names: Healaugh, Low Whita, and Thwaite in Swaledale; Hardraw, Swinith-
waite and Aysgarth in Wensleydale. All are characterized by their seventeenth-
century field barns which, isolated in the hay meadows, strongly contrast with
the compact longhouse of the Lake District where barn and farmhouse are
combined. The greystone Dales' barns provided shelter for livestock on the
ground floor with hay stored in the loft above. Their melancholy loneliness can
best be seen in Swaledale, especially at Gunnerside where the drystone fieldwalls
are also magnificently displayed. Andrew Young found that Swaledale '. . .
owes much to its walls . . . dragging up the fields and drawing down the fells, they
bind them in a unity. They cover the steep uneven slopes with their wavering
parallelograms.'

Wharfedale is different. It twists and turns, its valley floor is narrower, its field
barns less conspicuous and its woodlands more obtrusive on the scarred limestone slopes. But above all, in its lower reaches the river breaks through the
gritstone belt before emerging on to the industrial plain of the West Riding.
Thus, from Burnsall village to Ilkley, past the picturesque ruins of Bolton
Abbey, the fringing hills are darkly crowned with moor and tor. It is sombre
Millstone Grit that builds both the fieldwalls and the tiny villages in Lower
Wharfedale which, despite their attractive character, have lost the pervasive
glitter of the limestone. But Upper Wharfedale retains its pastoral stone-framed
charm: across from the cobbled streets of Grassington, for example, Linton is
one of the Pennines' most attractive villages. Its Norman church, handsome
Georgian houses, cottages, mills and ancient stone bridges cluster in perfect
harmony around the village green. No other settlement, perhaps, encapsulates
the ecclesiastic, domestic, agrarian and industrial history of the Dales' landscape
so well as Linton in Wharfedale.

# The English Stone Belt

Running in the shape of an ogee curve diagonally across the English lowlands
from Dorset to Humberside, a broad belt of limestone intervenes between the
sandstone country of the Midlands and the chalklands of the south-east. This line
of calcareous hills also extends a swathe of unspoiled countryside through the
heart of England and serves as a haven of agrarian tranquillity between the

industrial centres of Birmingham and London. It has been claimed that William Morris's anti-industrial philosophy, epitomized by the Arts and Crafts Movement nurtured in the Cotswolds, simply recognized the tenet that English culture is country-based. This romantic notion has led many people to see the central stone belt, from Somerset to Northamptonshire, as 'an ideal rustic world where life is uncomplicated by modernity'. It is certainly true that its scenery has a sense of permanence and timelessness due largely to the closely knit affinity between the physical and cultural landscapes. This relationship owes everything to the careful exploitation of the region's greatest asset, its local stone; from time immemorial there has been a resistance to the importation of exotic building materials. Consequently, the stone belt has retained a visual harmony in which 'villages resemble extended outcrops of the local quarry' and where J.B. Priestley found the walls '. . . still faintly warm and luminous as if they knew the trick of keeping the lost sunlight of centuries glimmering about them'.

The central English belt is made up of a number of bedded rocks that vary in colour, texture, thickness and porosity, laid down at times when clear tropical oceans dominated Britain's geological past, some 170 million years after the formation of the Mountain Limestone and 70 million years before the chalk. Intermediate both in age and geographical location, these Midland limestones also produce scenery midway between the bare, windswept uplands of the Mountain limestone and the lower, gentler downlands of the chalk. Except where they have been bent and fractured by the convulsions which threw up the Alps (and this is mainly along the Dorset coast), these mellow limestones remain undeformed. They lie in gently tilted sheets which descend imperceptibly south-eastwards until they disappear beneath the clays of south-eastern England. But their western and north-western limits are terminated by abrupt escarpments worn back by the incessant paring of primeval rivers. The Cotswold scarp at Broadway, however, appears to be implacable, dividing the Severn Vale from the vast rolling sweeps of cultivated wold and its scattered remnants of ancient sheep walk. Its 1,000 feet of extra height means that Cotswold wheat and barley ripen a full month later than in the vale. Because its stony beds are not quite so porous as those of the Mountain Limestone or the chalk, the rainwater tarries longer at the surface of the stone belt, its stream-beds are more replete, its woodlands more capacious, its chestnut-coloured soils more clinging. Yet, like those of the Mountain Limestone country, the rivers of the stone belt have bitten deeply into the well jointed rocks, fashioning steep-sided valleys in which ancient settlements cluster, sheltered in their stone-girt retreats. The colours of the lowland limestones range from the glittering white of Portland, through the buff and biscuit-coloured Bath stone to the honeyed gold of the Cotswolds and the gingery ochres of the South Midland marlstones. Additionally, there are the creamy-white dolomites of Yorkshire and the East Midlands but these are very much older and do not belong to the true stone belt. Each subtle tone is related to the varying amounts of iron oxide present in the rock; in fact the Northampton and Lincoln marlstones are so rich in iron that they are termed 'ironstones' and were once smelted in the steelworks at Corby and Scunthorpe.

The greatest difference between the upland limestones of the Pennines and those of the lowland stone belt is the dearth of bare rock exposures on the southern wolds. While wind, rain and ice have flayed the Pennines to expose their bony ribs, the lower elevations of the more southerly ridges remain discreetly clothed by vegetation which thrives on their deeper soils, unscathed by glaciers. Thick coverts of oak and hazel enrich the valleys, ash trees bestride the more exposed ridges while beech woods drape the larger scarps. The thicker soils have crept down the slopes, smoothing out the angularities of the bedded rocks and helped to create gently swelling profiles on even the steepest hillslopes. Only the Dorset sea cliffs and the Cotswold scarp at Leckhampton reveal their unclothed upright anatomy of stone and even there the rock faces are largely abandoned quarry walls.

In Dorset the limestones are restricted virtually to the appendages of Purbeck and Portland where they thrust their cream-and-white ledges into the breezy Channel to form such well known landmarks as Durdle Door and Stair Hole. While the so-called Isle of Purbeck, whose polished shelly stone was used inside Durham and Ely cathedrals, possesses some of the finest coastal scenery in England the same cannot be said for the Isle of Portland, 'a Peninsula carved by Time out of a single stone' (Hardy). Its pockmarked surface is a measure of the remarkable quality of the renowned Portland Stone, but almost all of it has gone into some of Britain's finest buildings leaving Portland itself in rather a sorry state. Its oolitic stone will be examined more fully on pp. 166–9.

Further north, where the stone belt wends through east Somerset and west Wiltshire and the villages relate happily to the rich farming landscapes, it intrudes a golden band between the grey Mountain Limestones of Mendip and the pale chalky pastures of Salisbury Plain. Its two famous quarries at Doulting and Chilmark have now ceased production but not before their building stones had left an indelible mark on the countryside. Doulting's fawn-coloured, shelly limestone can best be seen in Glastonbury Abbey or Wells cathedral's magnificent west front, though the humbler fourteenth-century tithe barns of this area are also notable. Chilmark Stone is a gritty form of Portland Stone, white when quarried but weathering to a greenish-grey, as illustrated by Wilton House and Salisbury Cathedral.

Stair Hole,
Dorset, where
Purbeck
Limestones
have been
crumpled by
earth
movements.

The stone belt achieves its apotheosis in the hilly tract of country which stretches for some 50 miles between Bath and Chipping Campden – the fabled scarpland of the Cotswolds 'where vernacular building in stone, flowered as nowhere else in Britain'. Rivers have carved the thick beds of limestone into broad swelling hills and deep winding valleys whose fast-flowing tributary streams were able to provide both power and a means of cleansing for the woollen industry, many of whose mills now stand empty around Stroud. The name Cotswolds, derived from Saxon, denoted 'Hills of the Sheepfolds', and it was the wealth derived from its ancient woollen industry that subsidized its memorable seventeenth-century churches, manors and cottages. One cannot quarrel with Jacquetta Hawkes's assertion that 'Men and sheep and the limestone hills have together made the Cotswolds realm.' Yet long before the woollen industry flourished in the Middle Ages or the Saxons had built their tiny church at Bradford-on-Avon the Romans had discovered that the Avon gorge had other

Drystone wall-building in the Cotswolds.

attractions. They founded their urban settlement of Aqua Sulis where a crustal fissure sliced through the oolitic rocks deeply enough to tap the boiling mineral waters of a buried igneous mass. From these antique beginnings the classical city of Bath has grown, climbing up the wooded slopes in a series of elegant Georgian terraces and crescents. The eighteenth-century masons had a plentiful supply of golden stone available from a dozen local quarries all of which are now closed save that of Monks Park near Corsham which, paradoxically, produces one of the poorest quality Bath stones. Although Bath's warm-toned oolite lends itself to ornamental work it does not stand up to the weather so well as its Portland counterpart. The Georgian builders were also faced with the problems induced by the steep terrain of the Avon gorge, for here the hard oolitic strata are underlain by softer beds of clay rocks which have been unable to bear their heavy burden. Several of Bath's surrounding slopes are scarred by landslips and some of its grandiose streets had to be terminated abruptly when they reached unstable ground. Nevertheless, Bath's magnificent golden architecture in its incomparable setting is a fitting termination to the southern limits of the Cotswolds.

For the next fifty miles one finds a chequerboard of russet fields and verdant woods; sheep-grazed pastures fragrant with thyme and stitched with drystone walls; clear streams with venerable stone bridges; mellow villages with homely mills and statuesque churches; drifts of cow parsley and elderflower on the roadside verges. Taken together these constantly recurring Cotswold themes create a landscape 'marked off by its peculiar genius from the outside everyday

An elegant town house, Stow-on-the Wold, Cotswolds.

world'. A close analysis will reveal, however, that it is not so much the natural elements but the stone buildings that make the dominant contribution to the Cotswold's scenic charm. Cotswold stone is certainly one of Britain's finest freestones, though the broken rubble, discarded by the quarryman, is always in perfect harmony when incorporated in the simpler cottage walls, especially when the dwellings are topped with local roofing 'slates'.

The majority of the traditional Cotswold buildings are roofed with slivers of hard flaggy limestone known as 'slate', though the rock was formed quite differently from the true slates of Wales. The chief source of supply was the small village of Stonesfield, near Woodstock, hence they are termed Stonesfield Slates, though the quarries have long been closed. During the autumn the honey-brown flagstones were hewn from underground galleries as large slabs of so-called 'pendle' and were then strewn on the surface to await splitting by the first hard winter frost. Once nature had done her work the villagers spent the

summer shaping and piercing the thin limestone sheets ready for distribution not only throughout the Cotswolds but also to roof most of the Oxford colleges. If the frosts were not severe enough to split the slabs the industry came to a standstill. When the slates were finally fixed on roofs they were carefully graded in size, with the smallest at the ridge and the largest at the eaves. The grading produces a satisfying perspective and the slates give a harmonious capping to the mellow stone walling, especially when the roofs become speckled with yellow lichens. In the villages south of a line from Cirencester to Burford other flaggy limestones, coarser than the Stonesfield Slates have also been used for roofing and occasionally, as at Filkins, for vertical stone fences. Many of these belong to the formation known as the Forest Marble, a term derived from the ancient Wychwood Forest, and they have the added advantage of not requiring frosts to split their thin flagstones.

Almost all Cotswold towns and villages are attractive, but some are outstandingly beautiful. The picture postcard village of Castle Combe, with its low-arched bridge, is archetypal and serves to start an aesthetic progression from the south. Next comes the elegant hillside town of Painswick, famed for its yew-shaded churchyard, then Bibury with its antique waterside row of cottages. The real show places have cathedral-like churches and perfectly preserved medieval streets of grey-gold houses: Burford, Broadway and Chipping Campden are the largest and best known; Bourton-on-the-Water, where the Windrush glides beneath tiny eighteenth-century stone bridges, one of the most perfectly contrived. But some of the lesser known and rather more reticent villages give an even great feeling of the true Cotswold way of life: Snowshill, high up on the scarp edge; the Barringtons, nestling by the Windrush, and particularly Lower Slaughter, whose trim little cottages beside greensward and stream belie the severity of its name.

It is something of a paradox that in his *Rural Rides*, published in 1830, William Cobbett took a more jaundiced view of the Cotswold scenery, for he was concerned not with the aesthetics of the stone belt but with its social welfare and agricultural productivity. He speaks of 'a very poor, dull and uninteresting country . . . divided into large fields by stone walls. Anything so ugly I have never seen before. The stone here lies very near to the surface. The plough is continually bringing it up, and thus, in general, come the means of making the walls . . . these stones are quite abominable'. Later he accurately describes the rich brown Cotswold soils as a 'stone brash', for they are liberally sprinkled with weathered flakes of stone. One of the bedded limestones is actually known as the 'cornbrash', a coarse, crumbly and lumpy stone which forms an indifferent building material, despite its widespread use in central Oxfordshire and North

(*Far left, above*) Arlington Row, Bibury, Gloucestershire, showing ancient roofs of Stonesfield Slates.

(*Left, above*) Cotswold flagstone fences, Filkins, Oxfordshire.

(*Left, below*) Stanton, a village built entirely in mellow Cotswold limestone.

Buckinghamshire. Its name is derived from the fine-quality corn which thrived on its well drained limestone soils. A similar coarse but tough stone, termed the 'Coral Rag', is another Cotswold limestone, composed of large shells and corals, from which many rubblestone buildings in and around Oxford have been constructed. The randomly uncoursed stones of the Coral Rag give its structures a sense of rugged permanence since it is hardly affected by weathering, which is more than can be said for another Oxford building stone, the infamous Heading-ton Stone. This attractive-looking freestone from local quarries was extensively used to construct seventeenth- and eighteenth-century Oxford, but with dis-astrous results. Jacquettà Hawkes describes how Renaissance architects, influ-enced by classical ideals, rushed to raise university façades of smooth ashlar. 'It was now that the intimacy between builders, quarrymen and stone broke down. Freestone began to be taken from Headington in great bulk and without the loving selection that went with the old understanding of the vices and virtues of every pit.' Thus, the poorly chosen stone began to blister and flake within a few decades, bringing a premature senility to the Oxford buildings which, though picturesque to such as Nathaniel Hawthorne, has led to centuries of costly restoration. Much of the replacement freestone has come from Taynton quarry near Burford, a source of good quality oolite since Anglo-Saxon times.

The River Evenlode is said to mark the eastern limits of the Cotswolds, with the golden-grey church towers of Gloucestershire and Oxfordshire being re-placed by the creamy-brown spires of Buckinghamshire and Northamptonshire. One recent writer believes that although in these latter counties 'stone villages abound, with angles and corners and textures so very nearly Cotswold, the valley settings, wooded hillsides and wind-washed western wolds have been left behind, and the landscape has lost its secret places.' Although a belt of oolitic freestones continues north-eastwards as far as Lincolnshire, once beyond Ox-fordshire the stone belt also includes another important limestone, the marl-stone. This ginger-coloured ironstone makes its greatest impact on the buildings of the south and east Midlands, though the contemporaneous Ham Hill Stone of Somerset (a type of shelly sandstone) has contributed to the vernacular architecture around Yeovil as well as in Montacute House. The marlstone is a rock layer older than the Cotswold oolite and in most places its bedded outcrop forms a stepped shelf at the foot of the oolitic escarpment, except north of Banbury where the marlstone creates the steep face of Edge Hill. The villages near Banbury display its ochreous colours to great advantage – Deddington, Adderbury, Aynho, Radway and Warmington – but it is the sage green and blue-grey stones from Hornton quarry that are most distinctive and these are the limestones sought by such famous sculptors as Henry Moore. In general,

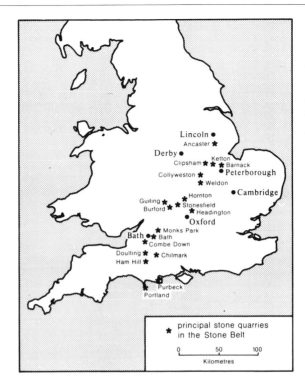

Figure 9
The major
limestone
quarries in the
Oolitic Stone
Belt.

marlstone creates a countryside of mixed farms with quickthorn hedges instead of drystone fieldwalls and in many of its buildings the rough textured brown marlstone is mixed with grey oolite to give a busy polychrome effect.

The lower oolitic scarplands of Northamptonshire and Rutland might have the textural quality of Cotswold freestone but they lack its scenic charm, partly because the land has been lowered by northern ice sheets and partly because its closely spaced market towns are rapidly becoming industrialized owing to their location midway between London and the Midlands. Canal, road and rail have allowed a rash of bricks and tiles to invade its towns and villages, except around Stamford, where the oolite produces another fine crop of stone villages. Moreover, the famous creamy-white freestones from Weldon, Clipsham, Ketton and Barnack have made as great an impact on buildings elsewhere in England as they have on the stone belt itself. Almost all the production from Weldon and Ketton, for example, was shipped by barge down the Welland and up the Ouse to build the glorious Cambridge colleges, where it has weathered considerably better than has Oxford's Headington Stone. Indeed, Oxford has recently turned to Clipsham Stone, the most famous of all calcareous freestones, following the examples already set by the Houses of Parliament and the cathedrals of York, Canterbury and Salisbury in their recent periods of restoration. York Minster was originally raised from nearby magnesian limestone, the creamy white texture of which was easy to carve and therefore popular with medieval masons. But more than any other stone this dolomitic limestone is severely affected by air pollution, leaving many stone churches around Wetherby and Pontefract in sad decay. Barnack Stone, from Northamptonshire, is no longer worked but its quarries functioned since Saxon times, supplying stone for the early Norman cathedrals at Peterborough, Norwich and Bury St Edmunds, the two latter belonging more truly to the chalklands. But, before turning to look at these, the last of the limestone landscapes, mention must be made of the Collyweston Slates of Northamptonshire, the eastern counterparts of the Stonesfield Slates. If Oxford is crowned with Stonesfield then Cambridge wears a diadem of Collyweston. Alec Clifton-Taylor prefers both the tone and the texture of the former while conceding that Collyweston Slates, because they split naturally into lighter and thinner sheets, produce a roof less heavy and therefore easier to lay. Owing to these important properties Collyweston roofs can be angled at slightly steeper gradients (up to 55°) than their Stonesfield counterparts. It is noteworthy that the heavy limestone roofing slabs used in the Yorkshire Dales result in low-pitched roofs (30° maximum slope).

## The chalklands

It has been suggested that there is nothing so quintessentially English as a chalk downland. Is this because the Wessex chalklands are seen as the cradle of early English civilization or because Stonehenge is one of the world's oldest and most illustrious monuments? Or is it simply because the downlands, once the home of countless sheep, unlock atavistic feelings of a long-lost rural domesticity based on mutton and wool? Perhaps it results from nothing more than an over-indulgence in reading the prose of Thomas Hardy, W.H. Hudson or Richard Jefferies, who remind us that until a mere fifty years ago the unploughed elemental simplicity of the turf-covered downs represented one of the least-changed landscapes of lowland Britain. Whatever the reason, the chalklands, although not landscapes of bare stone, have a very special place in the scenic heritage of Britain.

Chalk is not only the whitest and purest of the limestones but it is also the softest and most porous. One might be forgiven, therefore, for enquiring why the chalk can form hills like Walbury Hill in Berkshire, which rises to a height of 975 feet. There are two main reasons: first, despite its apparent uniformity, chalk possesses three relatively hard beds, all of which are resistant enough to have been used as building stones, as will be discussed in due course; secondly,

because of its porosity any rainfall sinks rapidly through the rock layers, leaving virtually no water available to aid river erosion at the surface. Unlike the massively bedded mountain limestone country, however, chalk lacks extensive networks of underground galleries and stream networks and instead has the properties of an enormous sponge. All the percolating groundwater eventually accumulates in the fissures and pores of the limestone until it forms a gigantic underground reservoir held on an impermeable base of underlying clay rocks. The saturated zone of rocks once provided the major water supply for London, since a layer of chalk underlies the thick beds of surface clay beneath the capital. Water was raised to the surface by pumps until the time when the reservoir became so depleted that London's water has now to be supplied almost entirely from the River Thames, although this too is now replenished from the chalk's upstream aquifers.

Chalk country. The South Downs escarpment near Alfriston, Sussex.

The top of the saturated zone, deep within the layers of porous rock, is termed the water-table, the height of which rises and falls according to the volume and seasonality of the rainfall. In most years the bulk of Britain's precipitation occurs during the winter, causing the hidden water-table to rise to such an extent that it reaches some of the deeper valley floors of the downlands. These are the places where seasonal streams, known as 'bournes', flow at the surface – hence the common occurrence of the name Winterbourne in the chalklands. During the drier summers, when the water-table falls, the ephemeral bourne-flow ceases and the villagers must rely once more on deep wells for their water supply. When the bournes do flow at the surface they are characterized by crystal clear waters, a measure not so much of the purity of the water but of the lack of clay sediments available for transportation in the streams.

Despite the generally waterless nature of the downlands, they are patterned by a remarkably integrated network of valleys, none of which is ever occupied by surface streams, not even bournes. These are the anomalous dry valleys of the chalk, phenomena which add a further dimension to the billowing hills and fluted hollows of the downlands. Although some major permanent rivers, such as the Thames and those of the Weald, rise on older rocks before crossing the surface outcrop of the chalk, it has long been difficult to explain how the extensive network of dry valleys came into being. Two simple explanations can be offered, both depending on changes of climate. The first necessitates a considerably higher rainfall which meant a permanently high water table and perennial rivers, possibly during the more tropical climates which prevailed several million years ago. Geologists tell us, however, that the dry valley network is much more recent and was probably carved out during the Ice Age when the subterranean water contained in the chalk's gigantic 'sponge' was permanently frozen into so-called 'permafrost', like that which still occurs in Arctic Canada. If the water from melting snow-patches on the downs was unable to soak down into the rocks it would have been forced to flow across the surface where its streams would soon have channelled the chalk. The resulting drainage network would not only have carved out the steep-sided valleys and coombes which diversify the gently swelling hills but would also have helped to break up a thick layer of sands and gravels which had become cemented by silica into a concrete-like carapace on most of the southern downlands. This armour-plated capping was a survival from Britain's pre-glacial tropical climates and was similar to the 'crusts' that today cap many hills in Africa. The tough sandy beds were broken up to form so-called 'sarsens' (derived from 'saracen', meaning a dark stranger), while the cemented gravels disintegrated into the boulders known as Hertfordshire Pudding Stones. Their scattered blocks, known as greywethers, can now be found strewn along

Sarsens littering a chalkland dry valley. Clatford Bottom, Marlborough Downs, Wiltshire.

many of the dry valleys and, in a countryside lacking good building stone, have been extensively utilized as corner stones, lintels and gateposts in the vernacular architecture. Yet it was not these tough sandstones which were to play the significant role in the evolution of settlement on the chalklands, for it was the presence of another hard material within the bedded chalk which was of much greater importance – that material was flint.

Flint has exercised more influence on early human history than almost any other stone in Europe, specifically because of its ability to be chipped into a variety of essential weapons and tools. Not only could axes, arrow heads and knife blades be fashioned from the durable stone, but it could also serve as scrapers, chisels, saws, picks, adzes and sickles in the earliest agrarian economy first developed by the Neolithic farmers. So valuable were the flints, which occur only in the upper layers of the chalk, that they were intensively mined from

Giant sarsens, Avebury, Wiltshire.

underground chambers on the South Downs of Sussex and at Grime's Graves in Norfolk. From there they were taken along ancient tracks or ridgeways on the grassy chalk ridges, high above the impenetrable forests of the clay vales, to service the requirements of the early downland settlers. In Wessex some of these primeval farmers tilled the thin, pallid soils of Salisbury Plain, which, because of its central location, became the focus of Britain's earliest organized society. From this node chalk ridges extend to the coasts of Norfolk, Kent, Sussex and Dorset. The wide, empty, rolling downland to the north of Salisbury is today a vast graveyard of prehistoric chieftains, a mosaic of ancient fields and a jigsaw of primitive village remnants. The burial chambers of Silbury Hill and the West Kennet long barrow are examples of the prodigious labour requirements of those times but even these major constructions are overshadowed by the monolithic stones of Avebury and Stonehenge which represent the culminating achievement of prehistoric man in Britain. The difficulties involved in manually transporting the gigantic sarsens from the downs and erecting them in ritualistic circles can only be imagined by a twentieth-century society with its mechanical earth movers. The story of the inner 'blue stones' at Stonehenge, laboriously shipped and dragged from far-off Pembrokeshire is almost impossible to conceive. At least Stonehenge has survived the fate which befell Avebury where some of the

40-ton stones were later to be broken up by the clergy (in an attempt to wipe out paganism) and subsequently by farmers for use as building stone; indeed some Avebury cottages have been constructed from irregular chunks of sarsen.

Once primitive man had discovered how metal tools could clear the scrub and forest of the lower valleys and vales the desiccated hill country was gradually abandoned and many centuries later the West Saxons, who gave their name to Wessex, were able to carve out their farmsteads on the valley floors. They settled mainly at crossing-points on the few permanent rivers that traversed the chalk-lands, as the Wiltshire names of Britford, Codford, Wilsford and Woodford clearly illustrate. While keeping their sheep traditionally on the magnificent turf of the grassy downland, the Saxons built their homes on gravel terraces in the sheltered valleys, where cattle grazed on the lush water meadows. Here too were the early water mills, grinding corn grown in small plots on the well drained valley sides, a far cry from the primitive stone querns of the earlier Neolithic farmers. Because of problems of water supply the astringent plateaux and hills of chalk were never settled again; henceforth they were to be left in calm serenity, populated only by hares, foxes, butterflies and skylarks.

Modern villagers and farmers live, as did their medieval forebears, in the same tiny hamlets nestling by the valley bournes or regimented along the scarp-foot springlines. Their traditional architecture depended primarily on chalkstone, flint and thatch until the invasion of brick and tile. The three hard beds in the chalk are known respectively as the Chalk Rock (youngest), the Melbourn Rock (used mainly in interiors) and the Totternhoe Stone (oldest), of which the latter is the most important building stone. In addition, on the Devon coast at Beer a hard chalky and shelly limestone, the Beer Stone, has been mined since Roman times but was most regularly used for fifteenth- and sixteenth-century churches in east Devon and west Dorset. The gritty Totternhoe Stone, quarried mainly on the Chilterns' escarpment, has survived in some of Britain's oldest buildings, such as Windsor Castle, the Dean's Yard at Westminster Abbey and the outer walls of the Tower of London. But, because of the chalk's lack of resistance to weathering, chalk building stone has been used mainly inside buildings, as in Ely cathedral's chapels. Externally, it endures longest when based on more resistant stone and is protected beneath overhanging eaves – 'give Chalk a good hat and shoes and it will serve you well'. That is why a few farmyards are contained within thatched walls of chalk blocks, as at Blewury, Oxon. Chalk blocks can be dressed and squared with relative ease and at places like Ashdown House and Uffington's Old School House on the Berkshire/Oxfordshire borders, it has been used as fine ashlar. Farther east, especially in East Anglia and Lincolnshire, its squared stones alternating with flint panels have been used to decorate many

*(Above)* Diaper work of flint and chalk, Wylie, Wiltshire.

*(Above right)* 'Give chalk a good hat and shoes and it will serve you well'. The thatched walls at Blewbury, Oxfordshire, are made partly of chalk blocks and chalky mud ('clunch').

churches in a chessboard design known as diaper work. East Anglia, devoid of building stone, has many flint-faced cottages and in north Norfolk whole villages are built of flint, with some flint church towers having been constructed in circular fashion because there were no cornerstones. Furthermore, because of its irregular angularity flint cannot be used for drystone walling so, unlike those of the stone belt, the chalkland fields are either hedged, fenced or left as open prairies. Where flint is used as a facing stone it is an excellent building material (so long as it is set in good quality mortar) because it is virtually indestructible by normal weathering processes. But because of its cold grey or sombre black appearance flint, whether 'knapped' (split) or used intact either as a pebble or as a 'horned' flint, is shown at its best when incorporated with the squared solidity of white stone or ruby brick. As a building material chalk stone is used much less frequently than flint. It lacks durability and is sparingly used for lintels, jambs

Tom Brown's School, Uffington, Oxfordshire. Finely dressed chalk blocks on a base of sarsens.

Brickwork and knapped flint is typical of downland architecture.

or corner stones, which means that, unlike Cotswold Stone, it never determines the personality of a town or village. The essential character of the chalklands, therefore, lies in the landscape itself.

From earliest times artists, writers and poets have been inspired by traditional downland scenery limitlessly extending beneath its wide, open skies. They never seem to have thought of it as being romantic, nor picturesque and certainly not as spectacular as the British uplands. Instead, they regarded the downs with homespun affection, they traversed their sheep-grazed expanses, marvelled at their cyclopean stone circles and depicted the solitude and the rustic simplicity of the downland life. In the seventeenth century Aubrey found 'the turf is of a good shorte sweet grasse, good for the sheep, and delightful to the eye', while even the over-critical Cobbett was delighted with the countryside of the Avon valley near Salisbury. The traditional unwooded chalk landscapes have a spaciousness, a sense of lightness and a delicacy of colouring, their pale soils barely covered by the springy turf. Their rolling skylines display so even a contour and so gentle a slope that eighteenth-century 'improvers' planted compact clumps of beech trees on some of the hill tops, simply to diversify the scene. Few have captured the atmosphere and the redolence of this antique land better than the pastoral poet Edward Thomas: 'The grass of the slope is mingled with small herbage, the salad burnet rosy-stemmed, the orange bird's foot trefoil, the purple thyme, the fine white flax, the delicatest golden hawk-bit, and basil and marjoram, and rosettes of crimson thistles, all sunny and warm and fragrant, glittering and glowing or melting into a summer haze, musical with grasshoppers and a-flutter with blue butterflies, so that the earth seems to be a thick-furred animal'. But the natural furry 'skin' of the downlands has almost disappeared, along with its once ubiquitous sheep, replaced by never-ending hedgeless fields of corn. These lands now perform as Britain's granaries, devoid of the ethos which made the grassy downlands unique. Yet, paradoxically, in winter, when the plough has turned the soil, brought up the broken flints and exposed the naked chalk on lean-soiled hillslopes, the downlands adopt a different guise. Whether it be Salisbury Plain or the Berkshire Downs, the bleached expanses beneath the cold void of the sky, unbroken save by flinty track or wind-bent hawthorn, display a wilderness as inexorable as many of the landscapes of stone described elsewhere in this book.

# THE SCENERY OF SANDSTONE

If chalk and some limestones bring cool tones to certain tracts of the British countryside, then the same cannot be said for most of its sandstones. Their vivid reddish hues are distinctive enough to have inspired early geologists into naming two of Britain's best-known geological formations as the Old Red Sandstone and the lighter-coloured New Red Sandstone, useful markers because these rubescent rocks were laid down just before and after the valuable coal-bearing strata. Their thick layers of red-brown stone were either deposited in shallow seas and lagoons adjacent to arid coastlines or were formed when desert dunes became cemented and fossilized by the passage of time. Jacquetta Hawkes perceptively describes the Old Red Sandstone as '. . . glowing with the remembered warmth of Devonian deserts' and the New Red Sandstone as possessing 'the same lingering glow of desert suns'. No one would dispute that wherever these particular rocks occur they infuse a cheerful rosiness to any scene, but these were not the only episodes in Britain's geological history when barren deserts prevailed and quartz sands clothed the land and choked the neighbouring seas. Some of Scotland's oldest rocks, the Torridonian Sandstones, have been eroded into some of Britain's most spectacular mountains. Moreover, many of the Pennines' most vaunted landscapes owe their appeal to the intractable Millstone Grits whose dark moorland scarps ring the Mountain Limestones and frown over the coalfields. Finally, there were temporary breaks in the deposition of the younger limestones of south-east England (that nature was later to fashion into the Cotswolds and the chalklands) during which times layers of sandstones, clays and shales were gently interleaved between the more massively bedded limestones. Today such rocks form the wide tracts of hilly country in the Weald and the North Yorkshire Moors.

Not all British sandstones are red; they display as great a variety of colours as an artist would choose for his palette: white, buff, ochre, yellow, orange, peach, pink, red, brown, purple, lavender, grey and even tints of green. The various tones depend, of course, on which particular minerals have stained the quartz grains. The purest sandstones, quite unadulterated, are pristine white or slightly buff, but the majority are imbued with different iron compounds that produce a miscellany of warmer colours ranging from yellow to brown. When clay minerals are present the sandstones take on a greyish tinge, while glauconite adds a hint of green to the otherwise bright red greensands.

Sandstones also vary in their hardness, hence in their resistance to weathering. Some of the toughest and coarsest are the conglomerates, in which primeval beaches of exotic shingle have been cemented into rocks as unyielding as concrete. Almost as obdurate are the finer-grained siliceous sandstones, so-called because their quartz grains are welded together by a tenacious bond of silica, the same material that makes flint virtually indestructible. The very tough sarsens of the chalklands are siliceous sandstones which explains why most masons found it impossible to fashion them into building blocks and used them instead as they were found. So too are most of the Pennine grits, but unlike the sarsens the gritstones could be dressed and used for a variety of purposes, not least as millstones. Moreover, their durable character is also resistant to polluted air, making them a valuable asset as a building stone in the smoke-blackened towns of the Pennine flanks. By contrast, the weakest sandstones are those bonded by less resistant clay minerals which allow the rock to weather fairly rapidly. Many British sandstones, however, are cemented by relatively resistant calcite which, though not as tough as silica, is certainly better than clay. Because the bonding material breaks down quicker than the constituent quartz grains it is this that determines the sandstone's strength, though a tough bond has its drawbacks. Building stone with a hard bonding material is marvellously durable but more expensive to quarry and much more laborious to carve. Nevertheless, using hard sandstone is more satisfactory in the long term for even if the less tenacious sandstones are cheaper and easier to work, their intricately decorated surfaces are the first to crumble, as exemplified by the tragic histories of Worcester and Chester cathedrals. Indeed, the great stone edifice at Worcester is invariably clad in a mesh of scaffolding and is one of three English medieval cathedrals built from New Red Sandstone (Carlisle and Lichfield are the others) that are constantly needing repairs. Paradoxically, both Cumbria and Staffordshire possess some of the finest New Red Sandstone freestones.

The glowing red and sultry brown rocks beneath the heart of England are composed almost entirely of New Red Sandstone. Occasionally, Coal Measure sandstones also bring a glimpse of warm stone where they have been quarried on coalfield hillsides to build such fine mansions as Keele Hall in North Staffordshire, but generally the New Red reigns unchallenged in the Midlands. Yet the latter rocks, despite infusing a glow to the rich Midland soils, make very little impact on the natural scenery of the English heartland. In general, sandstones weather into gravelly and sandy tracts, lacking the nutrients of the limestone regions and producing what the gardener calls 'well drained acid soils'. Consequently, many

Midland hills are clothed with heathland where pine and birch rise above wastes of gorse, bracken and heather. Only in the poorly drained valleys and plains, where solid rock is deeply buried, do the red soils support rich meadows or fields of arable crops. This agrarian patchwork includes copses of lime and oak, isolated survivals of the once ubiquitous broadleaf woodland felled by Saxon and Norman until only Needwood (Staffordshire), Sherwood (Nottinghamshire) and Wyre (Worcestershire) are left. It is true that some low sandstone ridges heave their bare rocky faces unexpectedly above the peaceful dairy herds browsing on the lush green meadows: Beeston Castle on its red precipice in mid-Cheshire always comes as a surprise, so does craggy Hawkstone Park in Shropshire; Kinver Edge cliff is a popular Staffordshire viewpoint and the rough hills of Cannock Chase form a favourite beauty spot though they are almost submerged by coniferous trees. But elsewhere the red stony bluffs have themselves been swamped by urban sprawl, spreading from such early defensible riverside cliffs as those at Bridgnorth, Warwick and Nottingham, which stood safely above the marshy floodplains. Thus, Midland England does not really qualify as a true landscape of stone. Its soils are too deep, its hills too wooded and its valleys too mantled with red-stained glacial sands and boulder clays to allow the rocky skeleton to show through. Cheshire, North Shropshire, Staffordshire, Worcestershire and Warwickshire are the truly 'red counties' of the Midlands, with Nottinghamshire as first reserve. Not

(*Above left*) Keele Hall, Staffordshire. A mansion built from local Coal Measure sandstones.

(*Above*) Beeston Castle, Cheshire, perched on its crag of New Red Sandstone.

Stokesay Castle, Shropshire. Welsh Border architecture frequently added half-timbering to a solid base of Old Red Sandstone.

surprisingly, their ancient oak forests and widespread brick clays have ensured that half-timbering became their earliest dominant building style. Nevertheless, even though English 'magpie' vernacular architecture reaches its zenith in the heart of England, its New Red Sandstones have still managed to produce a limited crop of stately homes and noble churches throughout the region, but particularly in Cheshire and Warwickshire. Like the sandstone cathedrals, however, the ornamental details of these attractive structures have crumbled away and have had to be replaced, usually by the pinkish-grey Hollington Stone, quarried near Uttoxeter, or by the red Grinshill Stone from near Shrewsbury. The latter town's historic core was originally built from Grinshill Stone, as was its predecessor, the nearby Roman Wroxeter, while Hollington Stone has served not only to patch the old cathedrals of Hereford and Worcester but also to create the new cathedral at Coventry. In Warwickshire, New Red Sandstone also makes a lasting visual impact in the military architecture of Kenilworth and Warwick but, in general, central England has been increasingly submerged beneath a red sea of a different

Worcester
Cathedral.
Despite its
elegance the
soft New Red
Sandstone
structure needs
constant repair.

kind, an ocean of brick and tile that disqualifies its rosy landscapes from a book describing Britain's landscapes of stone. The same strictures must also apply to the Greensands and Wealden sandstones of Kent, Surrey and Sussex which, despite the elevation of Leith Hill and the wild forested ridges of Ashdown, make a less significant contribution to landscapes of stone than do the encompassing down-lands of chalk. Reigate Stone and Bargate Stone may once have played a role in some of the Norman buildings in south-east England but their lime content is so high that they hardly qualify as true sandstones in any case. Moreover, the forests and clays of the Weald have produced their own wealth of half-timbering which again relegates sandstone to a subsidiary role in the cultural scene. One is left, therefore, with a mere handful of regions where bare sandstone rocks impose their presence indelibly on the countryside and these examples are widely spaced between Somerset and Shetland.

# Exmoor

Everyone has their own sentimental vision of Exmoor – if not from first-hand experience or the novel *Lorna Doone*, then from the pens of a variety of nineteenth-century poets such as Southey who wrote rhapsodically of its 'lofty hills which furze and fern embrown'. Yet some two hundred years earlier Exmoor was represented as little more than 'baren and morisch ground, where ys stone and breeding of yonge catelle, but little or no corne or habitation' (Leland). By 1725 Defoe was dismissing Exmoor as 'a melancholy view, being a vast tract of barren and desolate lands'. Such empty landscapes of fearsome wilderness remained repellent to public taste until the artists and writers of the Romantic Movement contrived to discover the picturesque appeal of 'Nature Untamed'. But even while they drew the first tourists to marvel at Exmoor's scenery, pragmatic 'improvers' waited in the wings ready to change the wild moorland face and create the cultivated landscape that prevails today. All because the soils beneath the Exmoor heaths, unlike those of Dartmoor, are deeper and potentially more fertile.

Modern Exmoor, rather more than most British moorlands, has something of a clinical look. Despite the broad stretches of heather, gorse and bracken on Dunkery Beacon, Winsford Hill and Withypool Hill, its natural starkness has almost disappeared, replaced by the cared-for appearance which reflects the land management associated with the dramatic reclamation of the early nineteenth century. Its flat moorland plateaux have always lacked the spectacle provided by the pinnacled tors which crown the rounded Dartmoor hills but now much of the untamed severity that characterizes Dartmoor has also gone. The scenic transformation was wrought mainly by a single family, the Knights, who from the early nineteenth century, having purchased the freehold from the Crown estates (for this had once been a famous royal hunting ground), set about implementing a remarkable phase of land 'improvement'. First, the impervious iron pan beneath the boggy peat was broken up to allow the surface water to drain freely away through the underlying porous rocks of the Old Red Sandstone. Once the newly drained sandy soils had been liberally dressed with lime to counteract their sourness they could be made reasonably productive. A lonely stone cottage at Simonsbath became the core of a flourishing village linked to the outside world by a network of lanes whose earthen shelter banks were planted with beech trees. Over a dozen farms were hacked out of the moor, their rugged sandstone buildings being roofed with Welsh slate. Such names as Pickedstones Farm and Duredown Farm reflect the difficulties of the reclamation and it is significant, moreover, that the only farms to have survived were those sited on warmer south-facing slopes in

Exmoor. Its moorland plateaux have largely been reclaimed and enclosed by hedges and walls.

the sheltered valleys among the windswept hills. Despite all the agrarian changes it is still possible to glimpse such moorland birds as the ring ouzel in the heathland and see the merlin circling overhead. The western end of Exmoor has the preponderance of its Red Deer but the wild ponies, said to be descendants of Bronze Age stock, graze ubiquitously throughout the National Park. Its most attractive villages cluster along the coast and in the eastern valleys, many of their stone-built houses having been colour-washed and thatched. Only in Dunster, with its substantial castle and much-photographed yarn market, does the natural sandstone peep through the whitewash, though villages such as Exford and Luccombe display varying degrees of bare building stone. The pinkish-brown stone of the tall round chimneys with their bulging bread ovens give the cottages a distinctive West Country air but, in general, sandstone does not dominate the architecture of Exmoor as much as granite commands the Dartmoor scene.

In the deeply carved valleys and the rounded desolate coombes of the gently rolling uplands, oak woodlands have flourished, imparting a luxuriance which belies the exposed nature of the heathery sandstone plateaux. Nonetheless, the level treeless skylines of central Exmoor do not generate dramatic landscapes – the scenic drama is reserved for its coastline. Down by the sea one discovers Exmoor's true landscape of stone, where tiny stone-built settlements crouch in thickly wooded bays hollowed out where less resistant slates intervene in the startling succession of mighty, sandstone headlands.

Some scientists believe that Exmoor's bulky northern shoulder is an example of a fault-line scarp, fashioned from the sandstone bastion left behind when the neighbouring rocks were lowered *en masse* into the Bristol Channel trough between parallel fracture lines, countless millions of years ago. Layer upon layer of purple grits, greyish shales and silvery slates, which comprise the Exmoor block, were gently tilted towards the south, leaving their severely truncated northern exposures to form the 1,000-foot escarpment overlooking the Bristol Channel. Although their slopes are not vertical they provide some of Britain's highest sea cliffs and unforgettable vistas that extend far northwards to the high Welsh massif. It was from Welsh and Scottish mountain masses that ice sheets were later to spread southwards across western Britain before being halted by the impregnable rampart of Exmoor. Streams from the melting ice front, unable to surmount the high sandstone scarp, were forced to run laterally along the face of its rocky wall, thereby carving a number of valleys which run disjointedly lengthwise along the coast between Minehead and Ilfracombe. Today, now that the ice sheets have gone and the melt-streams have been extinguished, many of these curious channels are perched high and dry on the coastal slopes, though some have been rediscovered by Exmoor's torrential streams. The most spectacular of the dry channels is known as the Valley of the Rocks at Lynmouth where a line of towering stony pinnacles stands between the empty channel and the plunging sea cliffs. These pinnacles are in fact tors, the jointed sandstones of Castle Rock and Ragged Jack producing rock piles as spectacular as any of their Dartmoor counterparts. Nevertheless, there are those who claim that these Exmoor coastal tors were probably produced not by deep tropical weathering but by prolific frost-shattering alongside the northern ice sheet when it impinged on this steep coastline.

Exmoor's north-flowing streams have had insufficient time to adjust their valley profiles to the precipitous northern slopes, which means that they still descend headlong in a flurry of white-water rapids and waterfalls, some of the latter plunging into the sea as vertical cataracts. Such energetic streams are capable of moving gigantic sandstone boulders from their upland reaches and down through the coastal gorges before dumping them bodily on the seashore. Many of those that form the stony delta at Lynmouth were created during one catastrophic storm in August 1952 when 9 inches of rain fell in 24 hours and the swollen East and West Lyn streams did their best to wipe Lynmouth off the map. It is difficult to conceive the immensity of these elemental forces when visiting the pretty stone villages of Lynmouth or Combe Martin, half-hidden in their thickly wooded hollows and apparently sheltered from the tempest. It is precisely because they stand on leeward coasts, facing away from the prevailing south-westerlies, that trees can flourish right down to the shore, especially between Lynmouth and Porlock and at

Woody Bay. Yet it is the bare coastal cliffs of Old Red Sandstone which leave the most lingering memories: the unyielding pyramids of the Great and Little Hangman, the jutting Foreland and the nagging immensity of the gradient up Countisbury Hill. The sensational northern façades of Exmoor more than compensate for the subdued and restrained moorlands of the interior. Its ragged coastal fringe, torn by torrents, buffeted by ocean waves, yet gently clothed with almost tropical verdure, is unique in Britain and is deservedly popular as one of its most beautiful coastlines.

Valley of the Rocks, Lynmouth, Devon.

# The Brecon Beacons and Black Mountains

The crowded industrial landscapes of South Wales's mining valleys give way northwards to deserted moorlands which in turn rise to the deeply dissected slopes of the Brecon Beacons and the Black Mountains. These high, bleak tablelands,

Pen-y-fan, Brecon Beacons.

channelled by the same rivers which slice through the coalfield, provide Britain's only example of a high mountain massif carved entirely from Old Red Sandstone, though most of Ireland's highest peaks are built from rocks of similar age and character. For some forty miles between Ammanford and Hay-on-Wye a gigantic sandstone rampart imposes itself implacably across the scene, standing back from the vale of the Towy but towering over the Usk and crowding the peaceful Wye.

Though broken by river valleys, this imposing wall of rock provides one of the most continuous exposures of bare stone to be found in southern Britain. It has been likened to a great wave, petrified on the point of breaking and, because all early geologists, influenced by the scriptures and the Biblical Flood, thought only in terms of marine agencies to explain landscape sculpture, its formation was long ascribed to ocean waves beating at the foot of a sandstone sea cliff. The discovery that its brooding, scalloped northern slopes were carved not by the sea but by glaciers was made only a century ago. Since then it has been shown how the flat-topped mountains would have accumulated thick masses of snow during the Ice Age and how the prevailing south-westerlies would have blown much of this off the high tops to accumulate on the shadowed north-facing slopes. Here it would have been transformed into glacier ice which would have plucked at the stony cliffs before streaming northwards to supplement the ice sheet filling the Usk's broad valley. Thus, many of the northern slopes were chiselled into amphitheatres backed by precipices whose frost-shattered rock flakes fell constantly from the receding plateau edge. Today, the amphitheatres are known as 'cwms' and some of their armchair-shaped floors are filled with tiny lakes such as Llyn-y-Fan Fach and Llyn Cwm Lwch. Their encircling red sandstone cliffs rise tier upon tier to the corniced summits, many of which surpass the 2,500-foot contour, but the rock faces are now moss-draped and scree-skirted where piles of frost-riven stones have accumulated in vast slope-foot aprons.

Only the western end of the mountainous escarpment has been sculpted by glaciers because snowfall was much heavier in the west than in the east, just as modern rainfall is heavier at the Welsh end of the ridge than at its English termination. The Old Red Sandstone massif can be divided into four parts: in the far west the Black Mountain (scaled by George Borrow and known colloquially as Carmarthen Fan) is divided from the heights of the Fforest Fawr by the headwaters of the river Tawe; in its turn the dark block of the Fforest Fawr is divided from its superior neighbours, the Brecon Beacons, by the infant river Taf; finally, the lofty Beacons are separated from the confusingly named Black Mountains, which overstep the English Border, by the broader valley of the Usk. Each of the mountain groups has its own distinctive character but all are unified by the intractable sandstone and its stark windswept moorlands. The table-topped masses bear a limited number of Welsh place names that not only occur time and time again but which, when translated, help to create a vivid word picture of this stony wilderness. The mountain summits are termed either 'pen' (=top) or 'fan' (=crest), diversified with 'carreg' (=stone) and 'creigiau' (=rocks); the slopes are dotted with names like 'rhos' (=moor), 'gwaun' (=mountain pasture) or 'rhedyn' (=bracken); and even the valleys are distinguished mainly by 'garw' (= rough), 'llwyn' (=shrub) and 'rhiw' (=slope). Considering the reddish colour of the sandstone soils the Welsh term for 'red' ('coch') is surprisingly rare, occurring most notably to describe the ancient fortress of Castell Coch, constructed from water-worn sandstone boulders. More commonly found is the term 'ddu' meaning black, probably because the sombre northern cliffs and summits appear dour and forbidding when seen silhouetted against the midday sun. Moreover, the black peaty mantle on their lofty plateau ridges carries a dark cover of bilberry and heather moor above the lighter tone of the grassy slopes. Here, on the higher tops, the kestrel and buzzard patrol the skies, while the heron, grebe and red-breasted merganser are the most common of the fish-eating birds which frequent the lower lakes and rivers.

It is not surprising that these desolate, windswept mountains held little attraction for the earliest settlers and it was not until the warmer climate which prevailed between 1800 and 600 BC that Bronze Age tribes brought their cattle and sheep to graze on their slopes. But apart from a few stone circles on the western hills there is little evidence of permanent settlement until the Iron Age when a

The scalloped escarpment of the Old Red Sandstone in South Wales: Black Mountain, Dyfed.

number of major hill forts were built on the outlying ridges and foothills. The forts were never permanently occupied but served rather as enclosures for livestock which grazed the mountain pasture, although the earthworks became refuges for the tribespeople in time of conflict. The hill forts are found predominantly on the flanks of the Usk valley between Brecon and Abergavenny, testifying to the age-old importance of the Usk's major breach in the sandstone massif as a means of communication. This idea of fortifying the Usk corridor was also adopted by the Romans and later by the Normans who built their great sandstone castles not only along the traditional routeway of the Usk valley but also to guard another invasion route along the northern slope of the Black Mountains between Clifford and Talgarth. The turbulent years of border warfare which followed the establishment of the Marcher Lordship in the eleventh century saw many of the surviving valley oakwoods destroyed, exacerbating the already treeless aspect of the mountains. The Great Forest of Brecknock, from which Fforest Fawr derives its name, was never a woodland in the modern sense but simply a title denoting that the moorlands were held as a royal hunting demesne. For this reason many of the hills have remained unfenced and ultimately the moorland became common land, with the old county of Brecon possessing more acreage of common than any other Welsh county. That of the Brecon Beacons has for centuries provided valuable supplementary grazing for the enclosed farmlands on the lower slopes. Here, in

Llanthony
Priory, Black
Mountains,
Gwent.

the valleys and foothills, the landscape is veined with a tracery of fieldwalls and hedgerows, in contrast to the stark unfenced moorland plateaux. Yet the high rainfall, lean soils and steep slopes have combined to make hill farming so difficult that many of the higher farmsteads have been abandoned. Furthermore, during the Industrial Revolution the thriving coal mines and iron works of the South Wales coalfield attracted many of the upland farmers away from their meagre rural livelihoods, leaving abandoned hill farms across the Welsh countryside.

Today, the mountains have additional functions, not always befitting their role as a National Park. Although their summits are a playground for walkers, their slopes have been swathed by Forestry Commission plantations and some of their valleys are dammed to provide water for the South Wales cities. But there are also hidden unspoilt valleys, thick with natural oak and dotted with mossy greystone ruins of woollen mill, farmhouse and monastic settlement. Llanthony Priory, for example, in the spectacular Vale of Ewyas in the Black Mountains, is deservedly popular as a beauty spot. Set amidst meadowland, below an avenue of beech trees planted by the poet Walter Savage Landor, it is the last sign of civilization on the twisting road up to the windswept Gospel Pass. From this high breach in the sandstone escarpment the Old Red Sandstone countryside of Herefordshire can be viewed as though it were a multi-coloured counterpane of fields quilted with woodland patches. Immediately below the precipitous crest is the prominent shelf of the Allt (= hillslope), grazed by countless wild ponies, and beneath that again the scattered stone farms, often whitewashed and sometimes roofed by russet brown flagstones. These are quarried from the lowest layer of Old Red Sandstone, aptly termed the Tilestone bed, which outcrops as a narrow band between Senny Bridge and Pont-ar-llechau (= bridge of the slabs). Wherever the plough has turned the soil of the foothills its redness matches that of the sleek Hereford cattle which graze the succulent water meadows. The soils are reasonably fertile where the 'Old Red' marls occur because they have a natural lime content. Partly for this reason, and partly due to religious beliefs, the red soils of a raw hillside scar were once scattered over the fields surrounding the sharp peak of Skirrid Fawr near Abergavenny. The locals believed that the 300-foot cleft on the mountainside could be attributed to the Crucifixion, but the truth is more prosaic; the red sandy earth was derived simply from a major landslide which left a livid gash on the sandstone slope. The most fertile tract of this upland border country is found in the eastern vale of the Black Mountain, the so-called Golden Valley, where legendary harvests of corn have been taken. Its fertility derives from the light loamy soils produced by the crumbly 'cornstones' (yet another different bed of the Old Red Sandstone) and partly from the infill of glacial drift, dumped by a glacier which widened the Golden Valley more than any of its neighbouring corridors.

A final word is necessary about the quality of the 'Old Red' as a building material. Although it is less friable than the New Red Sandstone it rarely yields a good freestone. The only working quarry left in England is in the Forest of Dean whose durable stones helped build the Anglican Cathedral at Liverpool. Some of its beds can be split quite easily so that in parts of the Welsh Border country it has been used for steps, paving and fieldwalls. In the main, however, like the New Red Sandstone, it has played a subsidiary role to the half-timbering of these parts, though where it has been used in the village churches and cottages '. . . its rough country face can nonetheless be very likeable and wholly appropriate to its surroundings', if we are to believe Alec Clifton-Taylor. By and large in many Border cottages the pinkish-grey stone forms only the foundation layers upon which the cruck-framed timbers have been erected to create a curious hybrid structure.

# The Pennine Gritstone

From Skipton and Ilkley in the north to Leek and Wirksworth in the south a darkly brooding moorland hoists its bulky frame into the rain-blurred skies of England's industrial north. Its celebrated backbone has gritstones for its vertebrae. Apart from the enclave of Mountain Limestone of the White Peak country, the uncompromising, darkly-weathered grit dominates the upland scene. From most high vantage points the skylines appear to be as sharply cut and as unadorned as the building stone itself, with the Peak National Park's numerous visitors able to testify to the unspoiled simplicity of the 'heath-clad showery hills' where peat bogs stain the brooks and 'silence dwells o'er flowerless moors'. Yet these apparently featureless moorlands are regularly trenched by swift-flowing streams and bounded by ice-scraped and frost-riven scarps which loom darkly over the valleyside villages and valleyfloor factory towns. The imprint of the Industrial Revolution is written large in these narrow dales for this is where moorland sheep and torrential streams provided the resource base for Yorkshire's famous woollen industry. This social and economic linkage between agrarian moor and industrial dale is captured visually in the landscape itself where gritstone fieldwalls descend steeply from the grassy tops or from the starkly silhouetted crags as if to bind the mills and terraced houses to the primordial rocks out of which their building stones were hewn. The austere but sturdy character of the Pennines' gritstone buildings produces a *genius loci* every bit as valid as that of the Cotswold limestone country but, unlike the villages of the Stone Belt where sheep brought prosperity as early as the Middle Ages, the accumulation of wealth from the Yorkshire woollen industry came a few centuries later. By that time the era of Renaissance rebuilding had almost run its course; great wool churches, such as those in East Anglia and the Cotswolds, were never to be built in northern England. When the affluence finally came to the Pennines their mills were often built in brick, their fashionable gritstone and sandstone houses (with a few notable exceptions) were of a more modest appearance and their humble stone cottages altogether more reticent than those of the Cotswolds. But these northern industrial towns, with gritstone quarries at every back door, were quick to seize the opportunities offered by this sturdy building stone. Nowhere in Britain can match the monolithic town halls and Gothic Revival churches that sprang up in the smoke-blackened Victorian towns which burrow into the bony Pennine hillsides.

Few British landscapes have been so comprehensively chronicled, in both fact and fiction, as those of the gritstone moors of northern England, and all descriptions are linked by a common theme, namely the severity of the weather

A derelict gritstone farm on Haworth Moor, West Yorkshire.

and the consequent bleakness of the scene. In the early nineteenth century Anne Brontë wrote of the wind's 'wild and lofty voice upon the pathless moor' while sister Emily described 'the mute bird sitting on the stone, the dank moss dripping from the wall, the thorn trees gaunt, the walls o'ergrown'. Charlotte singled out the 'bare masses of stone, with hardly enough earth in their clefts to nourish a stunted tree'. It would be difficult to find more concise and perceptive prose to encapsulate the ways in which the elements of wind and rain have fashioned the Pennine rock and regulated its vegetation.

The poorly drained expanses of the plateau tops are covered with extensive bogs where erosion has channelled the peat into yawning black gullies several feet deep. The lugubrious countenance of the moorland is barely lightened by the sprinkling of fluffy white tufts of hare's-tail cotton grass which dance constantly in the persistent wind. Except for a few patches among the tumbled rocks of the gritstone edges, gorse and heather are uncommon. Most surprising of all is the absence of the bright green sphagnum moss, a typical component of blanket bogs, something that can only be explained by the centuries of continuous grazing and periodic burning followed by the insidious effects of industrial pollution. Add to this the dearth of trees on the high moors, taken mainly for fuel or simply inhibited by pollution, then the gritstone plateaux today offer a prospect more stark than that seen by the Brontës some 150 years ago. Yet where the gritstone edges emerge from beneath the peat, the fretted rock creates a landscape that quickens the pulse and fascinates the eye. After the rolling monotony of the plateaux the ragged edges bring a diversity, a restlessness and an abruptness to the scene.

The black scarps of Stanage above Sheffield and of The Roaches behind Leek, which appear as craggy interjections in the smooth tablelands, will serve to illustrate those elements that make the Millstone Grit one of Britain's toughest sandstones. Although their well developed bedding and jointing has allowed frost, wind and rain to sunder their cliffs into a chaos of slabs, clefts and buttresses, the embattled edges of darkly weathered stone still stand proud as if to challenge the soot-blackened civic façades of the coalfield towns far below. Where the layers of tenacious gritstone have been gently tilted, like those at Ramshaw Rocks on the Leek-Buxton road, they project on the skyline like the prows of ancient warships. Elsewhere, especially on the moors above Hathersage, the flat-bedded rocks have been slowly whittled away to form tors, one of which was later to be fortified by early man into the brooding hill-fort known as Carl Wark. Like the granite tors of Dartmoor, the gritstone rock-towers of the Pennines come in all shapes and sizes but as weathering takes its toll on even these resistant rocks it is possible to discover the final scatter of boulders marking the last defiant remnant of a once mighty tor now disintegrated into sandy ruin. At the other extreme is the imposing turreted pyramid of Hen Cloud, a rugged isolated hill of gritstone, larger than a single tor, that has become naturally cut off from the The Roaches scarp by processes of erosion. Also detached from their adjoining cliff of bare rock are the spectacular gritstone towers known as Alport Castles, not far from the Sheffield-Glossop road. But in this case the pinnacled rock pile has been severed from its adjoining face not by normal erosion but by a gigantic landslip caused by the underlying layers of shale being unable to cope with their heavy burden of Millstone Grit. It is

The dark Millstone Grit of Stanage Edge, Peak District.

significant that many of the rock formations resemble buildings and have also been labelled with whimsical titles implying anthropogenic connections, none more so than the monolithic detached blocks of the Cow and Calf on Ilkley Moor. These particular rocks have, however, been artificially quarried from the natural cliff.

Many of the earliest gritstone quarries were opened specifically to produce grinding stones for use in such ancient water mills as that surviving at Nether Alderley in Cheshire, with the corn being brought from the surrounding lowlands. But today one sees millstones relegated to the role of roadside markers delimiting the Peak National Park. Sadly, these superb grindstones are no longer required commercially and the Millstone Edge quarry near Sheffield, which is littered with hundreds of half-finished millstones, has been described as a veritable graveyard of rural industry. Before the advent of the textile industry, the population of the Pennines would have been engaged almost entirely in livestock farming, for the cloudy uplands were no place for cereal growing. The earliest dwellings reflect the type of husbandry because the typical farmhouse evolved, like that of the Lake District, into a longhouse in which byre, barn and family abode were all incorporated beneath one lengthy roof. On the steepest slopes of the West Riding some of the longhouses were built end-on into the hillside, recalling the longhouses of Dartmoor. On the moorlands around Halifax and Huddersfield several eighteenth-century examples can still be seen. Their long squat outlines were very simply constructed from roughly squared masonry with virtually no ornament, largely because the earliest quarrying techniques were relatively crude and the tough grit was difficult to carve. In many cases the simple rectangular plan was broken by a projecting stairwell at the rear while the front usually had a large jutting slab (known as a 'pentice') to act as a snow canopy over the door. Only in the nineteenth century, when more sophisticated methods of quarrying, cutting and dressing had been introduced, were carefully shaped stones used regularly for window surrounds and door jambs. The sturdy gritstone dwellings, like those of the Yorkshire Dales farther north, were originally roofed with thick sandstone slabs, sometimes up to 4 feet in length. Though they are darker and heavier than the 'slates' of the Stone Belt, and certainly lack their elegance and refinement, the sandstone slabs do not require 'frosting', for they can be split easily by hammers. Nevertheless their great weight has caused many northern roof timbers to sag or collapse, leading to a widespread replacement of most of these traditional Pennine roofs by imported slate or tile. Although at one time there were several quarries producing the formidable inch-thick sandstone roofing slabs, those known in West Yorkshire as 'thackstones' were quarried almost exclusively at Elland near Leeds. The blue-grey flagstones commonly found on the older pavements and roofs of the

A Pennine farmhouse near Halifax, West Yorkshire.

east Cheshire and north Staffordshire hill towns often came from Kerridge quarry at Macclesfield, while the Lancashire side of the Pennines had its flagstones supplied from the Rossendale quarries.

In order to supplement the meagre income derived from upland farming many smallholders introduced weaving into their moorland living rooms as soon as the textile industry began to flourish at the end of the seventeenth century. As demand increased, new dwellings sprang up on the bleak Pennine slopes, most with specially constructed 'weavers' ' windows to allow light into the upper work-rooms, once the looms had been moved upstairs. Their window mullions were invariably of chiselled gritstone. The end of the eighteenth century, however, saw the demise of the cottage industry as steam replaced the water power of upland streams and the machine-looms of the valley-floor mills made moorland cottage production obsolete. Fortunately, many of these rugged moorland buildings have survived, a few reverting to agricultural uses but most having been converted for use as commuter homes. Some of the new factories and mills were built in brick, like that at Quarry Bank cotton mill at Styal in Cheshire, but there are a sufficient number of gritstone and Coal Measure sandstone buildings surviving in West Yorkshire and Derbyshire to convey a clear picture of the impact made by the Industrial Revolution on the Pennine landscapes. The tiny stone village of Cromford, for example, was built in a dale near Matlock by Richard Arkwright in the 1770s, where he was the first to use water power to drive cotton-spinning machinery. Hallowed as one of the sites from which the Industrial Revolution sprang, Cromford's surviving buildings of yellowish Darley Dale Grit have now

An isolated Pennine gritstone house near Littleborough, showing the first-floor weaver's windows.

been proclaimed a Conservation Area in which the mellow streets of workers' cottages combine with the mills and the canal wharf to produce a historic landscape of stone in miniature. On a larger scale, the planned woollen textile village of Saltaire, near Bradford, is equally impressive. Built by an early Victorian industrialist, Titus Salt, it demonstrates how improved quarrying and stone-cutting techniques could produce a perfectly trimmed and beautifully coursed urban fabric of sandstone and grit, an enlightened example of industrial town-planning in every sense.

A dichotomy can be identified between the unchanged rural villages of the isolated plateaux of the southern Peak District and those that eventually became part of the sooty heritage of the Pennines' industrial townscapes. Two of the least affected villages are perched high on the gritstone moors of north Staffordshire. Longnor, with its flagstone pavements and gritstone road setts overlooked by dark robust houses and a weatherbeaten church, has been described as 'a composition in cold stone'. Standing more than 1,500 feet above sea level, the tiny hamlet of Flash appears lost amidst some of the Pennines' barest and most intimidating scenery. Its stone-flagged roofs have been slurried with cement to withstand the gales which sweep the grimly-named Axe Edge where Pennine snow falls first and

lies longest. Though Stanton-in-Peak, near Bakewell, is less elevated and not so isolated, its eighteenth-century village plan of flagstone courtyards and gritstone alleyways has also remained unsullied by industry. On nearby Stanton Moor the prehistoric circle of the Nine Ladies, reputedly turned into stone for dancing on the Sabbath, recalls the antique landscapes of Cornwall where the Merry Maidens of Penwith and The Hurlers of Bodmin Moor bear similar legends. Not far away are the magnificent fine-grained gritstone mansions of Hardwick Hall and Chatsworth House, surrounded by beautiful woodlands. But these relatively unscathed landscapes of the southern Peak are exceptional and to the north of the flat-topped Kinderscout the upland villages, squeezed into the 'wasp waist' of the Pennines between industrial Yorkshire and Lancashire, are built from a coarser-grained gritstone 'whose native sobriety is soon deepened by a mourning veil of soot'.

Though the odd village like Mankinholes, astride its ancient packhorse route, appears to be relatively remote, most hilltop villages, such as austerely handsome Heptonstall, usually look down on the industrial chimneys of valley towns like Marsden, Todmorden and Hebden Bridge. Nonetheless, a modern author, Geoffrey Wright, sees a different aspect to these moorland towns and 'their

Chatsworth House, Derbyshire.

Heptonstall, West Yorkshire: an industrial landscape 'touched with magic'.

huddles of houses with dark Victorian churches and darker mills', for he believes that when '. . . sunlight breaks through to glisten on shiny roof tops or gleam on bright green fields, throwing up in harsh relief the wandering walls, then are these northern industrial landscapes touched with magic.' Emily Brontë, writing from the darkly brooding parsonage of gritstone Haworth, that 'last outpost of the inhabited West Riding', expressed similar optimistic sentiments. She found the summer moorland inspirational and alive with birdsong, 'awake and wild with joy', where there was '. . . nothing more divine than those glens shut in by hills, and those bluff, bold swells of heath'.

## The North York Moors

To a lot of people the North York Moors are off the beaten track, by-passed en route to Scotland and largely unexplored by anyone who lives very far beyond the realms of Yorkshire and Teesside. Yet this northern limit of England's great Stone Belt has a wonderful panoply of unspoiled moorland scenery and a coastline to rival any in southern England. Its relative obscurity is borne out by the dearth of topographic description: dismissed almost without comment by Defoe, unvisited by Cobbett and ignored even by J.B. Priestley, one of Yorkshire's most discerning writers. Unlike the gritstone moors of the Pennines, the austere heaths of the North York Moors have failed to produce either a major regional novelist or a poet (if one excludes Whitby's monastic Caedmon of Anglo-Saxon times). Yet at the turn of the century the tiny fishing village of Staithes had an artist's colony that rivalled the eminence of Newlyn in Cornwall. Two decades later Edmund Vale was writing that 'Every bit of Staithes is picturesque, its shop fronts, its river harbour, its waterfront whereon stands the Cod and Lobster Inn, tarred all over, wall and roof.' Tourists have subsequently discovered that the charms of such villages, their coastal cliffs and their inland dales are due in no small measure to the character of the rocks of these parts.

It may come as no surprise to learn that the fine-grained yellowish-grey and brown sandstones, which cap the tableland of the North York Moors, are of the same age as the oolitic limestones of the Cotswolds. Though these northern sandstones are underlain by countless layers of bedded grits, shales and limestones, it is the 200-foot capping of golden rock which has produced the rolling moorland and given the tawny glow to its villages and towns. The sandy rocks, like those of the much older Millstone Grit, were laid down in a vast delta, which extended into the gulf of a tropical sea cut off from the clear ocean in which the oolitic sediments were being deposited farther south. Because many of the ever-shifting river channels, which deposited the deltaic sands, were tidal, they became filled with mud banks on which transient swampy forests were later to live and die. In this way the russet-coloured sandstones became interbedded with blue-grey shales and thin organic layers, themselves compressed into narrow seams of coal or wisps of hard black stone known as jet. No sooner had the bulk of this sedimentary 'layer cake' been uplifted into a gently inclined dome than voracious rivers set about carving the multi-coloured confection into numerous valleys and the insatiable ocean began gnawing at its eastern limits. Between them, the forces of nature have reduced the pile of bedded rocks to a 1,200-foot high plateauland, steeply scarped to the west and north, wave-truncated in the east and deeply engraved by dales throughout its centre. Only in the south do the moorland slopes and valleys descend in comparatively easy gradients to the broad flat floor of the Vale of Pickering and, not surprisingly, this is the tract where most of the larger settlements have been sited.

When prehistoric man first came this way he would have found the sandy hilltops lightly wooded in contrast to the virtually impenetrable forests of the marshy vales. Thus, the easily worked upland soils in the woodland clearings

Fat Betty, an ancient waymarker on the North Yorkshire Moors.

proved attractive enough to provoke a substantial primitive colonization of these eastern hills, especially since water supply was no problem. By the close of the Bronze Age, when the burial mounds had already begun to intrude their earthy humps into the plateau skylines, most of the upland trees would already have been felled. And because of the deteriorating climate and the fact that the soils had been impoverished by overgrazing and careless cultivation, the woodlands were never to return. When the Romans came to northern England they discovered that these sandstone tablelands had been transformed into desolate heather moors, although thickets of ash and oak had survived in the deeply etched dales. The legionaries found that the local sandstones helped them in their road making and some stretches of these stony ways are still to be seen. Other tracks were later to be followed as packhorse routes, picked out by a variety of stone crosses which assisted early tradesmen to find their way across the moors through hill fog and sea 'fret'. Today, a mere handful of the crosses survive, including the delightfully named Fat Betty. The final invaders were the Saxons and the Norsemen who settled in the valleys and built their villages on the steep coastal bluffs. It was the Vikings who left their mark on the place names, for the moorland hills are known as 'brows', the streams as 'becks', the coastal headlands as 'nabs', which were themselves divided by deeply cut coastal embayments termed 'wykes'. But the few inland villages that are found on the moorlands are almost all hidden in the dales, leaving the broad heathery ridges ('riggs') populated only by red grouse, curlew and lapwing.

The valleys are surprisingly deep and gorge-like in places, their slopes rising up

Rievaulx Abbey, North Yorkshire Moors.

to perpendicular walls of yellowish sandstone. Elsewhere the hard rock bands stand out as valley side shelves giving a stepped profile down which tributary becks tumble in great leaping waterfalls before disappearing into riverbank woodlands carpeted with bluebells, anemones and daffodils. Here, in this damper, shaded, bosky environment, the badger scuttles, the squirrel bounds and the woodpecker glides.

Not all of the dales are as wooded as those of Rosedale and Ryedale where the beautiful golden sandstone ruins of Rievaulx Abbey adorn the valley floor. The northern valleys of Eskdale, Westerdale and Glaisdale, for example, are almost devoid of trees save those that shelter farms and pick out the field boundaries. Most spectacular of all, however, is the serpent-like Newton Dale which gouges through the entire breadth of the eastern moorlands between Goathland and Pickering. Not only is it more canyon-like than the other dales but its 500-foot deep trench is virtually streamless, posing a question as to its formation. It has been suggested that its gorge was carved out rapidly by the same processes which created the Valley of the Rocks on Exmoor, namely the ephemeral floodwaters from a melting ice sheet. In this case the ice front stood across the northern hills and blocked the seaward exit of the Esk valley, thereby causing the melting waters to seek an alternative route southwards. Today the ravine has provided a fortuitous

Boulby Cliff,
near Staithes,
North
Yorkshire.

throughway for a railway, even though roads prefer to keep traditionally to their
ridge tops. Not far from the Pickering-Whitby road, just to the east of Newton
Dale, a layer of hard gritstone, which underlies the sandstone beds, has been
sculpted by wind and frost into the spectacular tors of the Bridestones. Their
mushroom shapes and pedestal forms can be understood if one notes how the less

resistant limestone beds below the grit have been worn away more rapidly than the hard caprock.

Different degrees of resistance to weathering and erosion have meant that the coastal rocks also exhibit a wide variety of landforms where the moorlands are abruptly truncated by the steely grey waters of the North Sea. In general the high headlands are built from layer upon layer of hard sandstones and limestones while the clefts and wykes have been carved where softer shales predominate. The broader bays, such as Robin Hood's Bay and Runswick Bay, are plugged with glacial boulder clay, so that their curved rims are fringed with sandstone sea cliffs only at their extremities. Where the coastal streams have cut narrow ravines down to the rocky shore they provide the only means of access to the interior from this rockbound coast. Not surprisingly, the coastal villages have had to be sited in these narrow gaps, in much the same way as those of north Devon and Somerset, which they closely resemble. The village of Staithes, in a cleft more imposing than that at Whitby, is made more spectacular by the backdrop of Boulby Cliff, at 672 feet the highest of England's vertical sea cliffs. But the ancient port of Whitby is one of England's most memorable towns. Nestling below the river cliffs of the Esk, it has managed to preserve its medieval identity of mellow sandstone buildings topped with scarlet pantile roofs. It is true that bricks also abound but it is the yellowish-grey sandstone from the nearby Aislaby quarry that gives Whitby the air of a stone-built town in a stone-girt setting. The brown stone quays, the greyish stone street cobbles and the yellow-walled cottages of The Yards combine to make Whitby England's archetypal sandstone fishing village. The twisting flight of 199 sandstone steps leading up to the cliff-top church of St Mary helps to give Whitby such an unforgettable townscape. The gaunt and severely weathered headstones of the churchyard graves are overlooked by the squat Norman tower, which is overshadowed in turn by the haunting silhouette of the seventh-century abbey ruins, all built of the golden-grey Aislaby sandstone.

# Northern Scotland

A dichotomy exists between the sandstone lands of Northern Scotland, a division which in part reflects the different ages and character of the sandstones themselves. On the western seaboard the landscapes between Applecross and Cape Wrath are dominated by the overwhelming mountains of Torridonian Sandstone whose rocks were laid down more than 600 million years ago. In contrast, the eastern coastland, from Inverness to Orkney, is the realm of the Old Red

Sandstone, the sediments of which are only half as old as those of the Torridonian. To the layman it must be something of a paradox to find that, despite the considerably longer period of weathering and erosion endured by the older rocks, it is they that build the high mountainlands, some of the most spectacular in Britain, while the more 'youthful' Old Red rocks of the east create mere plains and coastal platforms.

North-west Scotland has some of the wildest and most barren tracts of the British countryside, a landscape of limitless salmon-coloured rock regularly draped with blanket bog but only sporadically patched with tattered forests of native birch and pine. It is a wilderness of few settlements and with a handful of single-track roads that deter all but the most avid naturalists or those who seek solitude off the beaten track. The north-east corner of Scotland, by contrast, where thick layers of Old Red Sandstone fringe the North Sea, has an additional veneer of twentieth-century development, for its gentle coastlands have a network of railways and main roads serving handsome stone-built towns set amidst a chequerboard of farmland that thrives on the relatively fertile soils.

An explanation of the contrasts can be found by examining not only the contrasting nature of the sandstones themselves but also by understanding the different environments that prevail, respectively, along the exposed Atlantic shoreline and on the sheltered eastern coasts. The Torridonian Sandstone is composed largely of almost indestructible conglomerates and siliceous gritstones, varying in colour from anaemic pink to strident red. Its coarsely grained rocks, while withstanding the weather almost as successfully as granite, are generally too hard to quarry and thus to be used successfully for anything other than a crude local building material. The few crofting cottages in which Torridonian rock has been utilized are characterized by their chunky, unsophisticated stonework, profusely covered with yellow and grey lichens except where they have been limewashed. The roughly built structures often incorporate glacial boulders and pebbles taken from storm beach or stream bed, thereby maintaining the close visual affinity between homestead and landscape, a relationship all the more striking in such primitive surroundings as those of north-west Scotland. One senses that the traditional whitewashing of the cottages must be more of an attempt to brighten the dour masonry, ensconced as it is in a desolation of peat bog, than to protect the impermeable sandstone from the stormy climate. Yet wherever the slabby bedrock projects from beneath its shroud of peat the sandstone's rosy tones introduce an unmistakable glow to these cold northern latitudes.

Eventually, after aeons of weathering, even this coarse, pebbly rock will disintegrate into a scatter of chaotic gravel, leaving a sandy residue with virtually no potential fertility. Moreover, it is hardly surprising that in such an ice-scoured

terrain as this the few rudimentary soils that were created have subsequently been swept away, leaving little more than glacially polished slabs to reflect the windy sky. Torridonian country, therefore, is characterized by its primordial nudity whose starkness is exaggerated by the flesh-coloured tones of the sandstone. It is also noted for the grandeur of its monolithic peaks; its thickly bedded rocks build some of Scotland's most striking summits. The latticework of bedding-planes and joints has been exploited by the elements to produce a ragged profile of pinnacles, turrets and pyramids, more reminiscent of Mayan temples rising above tropical swampy plains than of the windswept sub-Arctic blanket bogs of northern Scotland. Such ice-gouged fjords as Loch Torridon soon dispel any such fantasies wherever they bite deeply into the mountainland. Golden eagles and buzzards wheel slowly above the mighty precipices, while ptarmigan and snow-bunting nest on the stony ridges amidst the Alpine chill. Below, in the shaded solitude of the pine forests that creep up into Beinn Eighe's northern corries, the rarely observed wild cat and pine-marten skulk between the moss-covered rocks. But the towering amphitheatres of florid stone stand hushed and empty, their silence broken only by the occasional bellow of Red Deer stags.

Despite the bulky spectacle of the Torridon Highlands and the citadel of An Teallach, to witness Torridonian Sandstone country at its most primeval one has to visit the unforgettable glacier-scarred giants of Assynt whose soaring buttresses

Loch Torridon, western Scotland, carved by ice from ancient sandstones.

The stately cone of Stac Polly, western Scotland.

and towers are unmatched in verticality anywhere in Britain. While some peaks like Canisp are little more than stately cones and the castellated turrets of Stac Polly may be matched by other Scottish summits, the isolated spire of Suilven is unique, for it stands like a solitary sandstone lighthouse above an alien sea of ice-scrubbed metamorphic rock. Assynt's pink mountains of bedded stone are all flagrantly bare, their rocky screes are only thinly stitched with moss, the denuded platforms from which they rise carry scarcely a blade of grass and not a branch of heather – this lunar scene is probably the starkest of all our landscapes of stone.

The scenic disparity between the Torridonian terrain of the west and the Old Red Sandstone country of the east could not be more striking. It is true that the North Sea coastlands also have their peat bogs but there the rainfall is lower, the wind less strong, the rock less obdurate and the land less broken. Above all the Ice Age glaciers have been less rapacious, for the eastern plains have gained much of the ice-transported debris that was chiselled from the western mountainland. Though the boulder clay and sands washed from melting ice sheets do not in themselves create highly productive soils, at least they provide a more receptive medium for agriculture than the unyielding rocks of the west. Nonetheless, since the Old Red Sandstone is almost everywhere thickly blanketed by glacial drift and peat, it can make its major scenic contribution only at the coast.

With its endless tracts of bog, the interior of Caithness is as uninspiring a place as any in Britain. Yet where the mundane moor is suddenly halted by the stormy waters of the Pentland Firth and the ruthless North Sea waves its underlying sandstone springs to life in the rubescent precipice of Dunnet Head and the imposing sea stacks of Duncansby. Between these two rocky projections lies the famous extremity of John O'Groats where travellers once embarked for the neighbouring Isles of Orkney.

The northern archipelagos of Orkney and Shetland exhibit a variety of moods according to the character of the weather. In sunshine their scenery is hauntingly beautiful, a jewel-like mosaic of rocks and skerries encompassed by the immensity of the sky; a shifting dapple of sun and cloud shadow across a creamy fringe of surf on cliff-girt shores. In adverse weather, however, their wind-seared, treeless hills appear merely stark and gaunt, ravaged by Atlantic waves and Arctic storms. Yet despite their northerly latitude, where a short growing season inhibits certain farming practices, these remote islands are by no means barren and infertile. Above all, the Old Red Sandstone has given to the Orkneys, and to a lesser extent to the Shetlands, a wonderful building stone which makes its presence felt wherever the human imprint is found in the landscape.

No description of Orkney is complete without reference to the wind for it invades every aspect of Orcadian life. Because of the incessant windiness Orcadian woodlands are rare, so are hedgerow trees. A recent author has commented that 'one usually thinks of trees sheltering houses; in Orkney houses shelter trees and the largest trees on the islands are to be found in central Kirkwall'. In fact the main thoroughfare of Orkney's capital has a large tree right in the middle of its flagstoned street, though the majority of the trees are found around the sandstone ruins of the Earl's and Bishop's Palaces. Overshadowing these sandstone buildings and their flagstone streets is Kirkwall's redoubtable cathedral of St Magnus, a twelfth-century Norman structure reminiscent of that at Durham. Built from blocks of pink and ochre sandstone, brought from a neighbouring coastal quarry, this simple but imposing edifice recalls the early Scandinavian links, for it was not until 1468 that Orkney and Shetland became subject to the Scottish crown. But the Norsemen were not the first to make use of the slabby sandstones as a building material; prehistoric settlers had built structures that were already thousands of years old when Viking invaders scratched runic graffiti on their walls.

The Old Red Sandstone of Caithness, Orkney and Shetland is much more finely grained and evenly bedded than that of Exmoor and South Wales and contains many beds of easily split flagstone. This invaluable building stone has been in constant use since the Stone Age to construct such diverse features as pavements, roads, drystone walls, roofing 'slates', gravestones and, more singularly, when

The sandstone splendour of St Magnus' Cathedral, Kirkwall, Orkney.

stood on end, the gawky flagstone fences of these parts. The bedded flags recline naturally at such easy angles that where they are exposed along the coasts the ocean waves have greatly assisted mankind by completing much of the initial quarrying. It is likely that the henge known as the Ring of Brodgar, between Kirkwall and Stromness, had its 15-foot stones simply dragged from the nearby shore. This may also have been true in the equally dramatic Stones of Stenness which form an 'avenue' reminiscent of that at Avebury. The neighbouring island of Hoy (Norse='High Island'), whose name reflects the presence of a coarser, blocky sandstone from which its higher hills and beetling sea cliffs have been carved, can claim to have two superlative features unique in the whole of Britain. The first is a remarkable sea stack, the Old Man of Hoy, a pillar of bedded red sandstone higher than any other along the entire British coastline. The second is a colossal ice-dumped boulder whose centre has been hollowed out by Stone Age man to form the only example in Britain of a rock-cut tomb similar to those of Mycaenean Greece. No less striking is the cyclopean passage-grave of Maes Howe built some 4500 years ago. Though the stonework is covered by a huge dome of earth, its drystone slabs fit so well together that it is impossible to insert a knife blade between them. There seems little doubt that the naturally smooth character of the flagstones played an important role in the final quality of this masonry. The same is true of the even more remarkable prehistoric village of Skara Brae, lying

Old Red Sandstone sea cliffs and the former crofts of Rackwick village, Hoy, Orkney.

The prehistoric
village of Skara
Brae, Orkney,
showing
ingenious use of
flagstones.

The broch on
Mousa Island,
Shetland.

A flagstone
lane,
Stromness,
Orkney.

some 5 miles west of Maes Howe. Here, on a gale-lashed coast, an entire Stone
Age village was buried by a sandstorm about 3,000 years ago. Not until it was
revealed by another storm just over a century ago were archaeologists able to
reconstruct the life style of these early Orcadians. They were able to show that
cattle and sheep were reared and that shellfish were eaten, but more significantly,
in view of the readily available flagstones, that almost every item of furniture and
tableware was fashioned from stone, to say nothing of the flagstone walls and
floors of the huts themselves. Skara Brae, together with Shetland's stone village of
Jarlshof, make up Scotland's most fascinating prehistoric settlements, but while
Skara Brae's Neolithic culture became 'fossilized' by the disastrous sandstorm,
the settlement of Jarlshof is known to have been occupied almost continuously
since it was built in Neolithic times: Shetlanders lived in Jarlshof's flagstone
dwellings not only through the Bronze Age, Iron Age and during the Viking
settlements, but also through medieval times and on until the sixteenth century.

Today the Orkney farms, with their neat stone buildings and flagstone fences,
carry on the age-old tradition of utilizing the geological resource given to them by
the Old Red Sandstone. Their roofs, burdened with weighty flags, are pitched at

low angles, like those of the gritstone Pennines, but regrettably (though under-
standably) many 'slates' have been replaced by characterless asbestos tiles. On the
Orcadian main island the green pasturelands are still crowded with cattle etched
against an endless background of sea and sky for here the landscapes have wide
horizons. So too did the sailors who once called at their ports en route to the Arctic
fishing grounds. Sir John Franklin's last British landfall was Stromness ('The
Haven inside the Bay') before he set out on his fateful voyage to the North West
Passage. He would recognize the virtually unchanged Scandinavian-style port
tucked against the hillside, its houses still jostling shoulders along the stone-
flagged street and pushing their crow-stepped gables into the sea. Lerwick,
Shetland's capital, was once like Stromness, for its Old Red flagstones and
sandstones originally produced a pleasing and wholesome townscape around the
harbour. But unlike the curiously stone-built hollow towers ('brochs') at Clickhi-
min and on nearby Mousa Island, Lerwick's antique stonework has been largely
overwhelmed by a rash of modern development associated with the North Sea Oil
bonanza.

# SLATE IN THE LANDSCAPE

Although some limestone and sandstone roofing slabs are colloquially referred to as 'slates', true slate is formed only when fine-grained rocks, such as mudstones and beds of volcanic ash, are squeezed by crustal convulsions or baked by igneous rocks. The pressure and heat of these metamorphic episodes not only close the pore spaces to make slate the least porous of rocks but also create a completely new texture, termed 'cleavage', which allows the stone to be split across the original bedding-planes. Since slate has such excellent waterproof qualities, both as a roofing material and as a building stone, it is paradoxical that when left in its natural state, unquarried on a hillslope, slate produces not the tough unyielding crag that one might expect but rather a smoothly rounded mountain shoulder. In fact, wherever slate is found in upland Britain it creates a featureless rolling terrain, its benign influence being especially noticeable where it rubs shoulders with the bulky volcanic rocks of Cumbria and Snowdonia whose rugged peaks and jagged crags make the neighbouring slate country seem comparatively tame. Yet a closer look will reveal that where slaty rocks are attacked by the weather their exposed faces are not so smooth as they first appear. When examined in detail the bare slate is found to be delicately splintered. The stone which splits so thinly and evenly when skilfully struck with hammer and chisel also yields to nature's incessant attack. Rain, frost and wind have etched the surface of the stone as surely as the engraver's acid bites into a copper plate. This is particularly apparent where the smoothly textured rock stands on end and has been slowly weathered into a minutely serrated surface whose razor-sharp slivers are kept as tightly packed as the pages of a closed book. But as pressure is released over aeons of time the thin wafers of slate break away from the bedrock and slide down the hillsides to form voluminous aprons of scree. Unable to penetrate the scree's impervious slices of rock, rainfall runs off the 'scaly' slopes with some rapidity, lubricating the scree and making it notoriously unstable. After every prolonged storm the hillsides appear to shiver as broken slates glissade down to the valley floor. Because of this instability it is not surprising to find that both natural screes and mounds of quarry waste are almost devoid of soil and therefore of vegetation. The scenery around slate towns, therefore, is barren and austere; landscapes of bare, broken stone are so dominant that one's senses are bludgeoned into a feeling of utter desolation.

A slaty track in
Snowdonia.

The scenery surrounding the North Wales slate quarries, for example, was adjudged to be so disfigured that when the Snowdonia National Park was designated in 1951 the commission excluded all the working quarries.

It is an interesting commentary on current landscape tastes that the public is said to regard quarries as aesthetically unacceptable. Yet many of Britain's slate quarries stand next to glacially excavated hollows of comparable stature and form, the crude sculpturing of which is held in very high regard as part of our mountain heritage. Perhaps the reason is not simply the visual dereliction of the slate workings but also intangible environmental considerations such as the relatively poor quality of life and working conditions in the slate towns. The melancholy greyness of their building stones beneath the mountain mist and rain does little to boost one's value judgement, yet the same weather conditions in the nearby glacially scoured corrie have often been described as adding to the majesty of the view. Is it simply because the regularly spaced working faces and spoil heaps bring an unnatural dimension to the upland scene? Glaciers rarely produce such layered symmetry in their rocky amphitheatres. But once more this cannot be the entire explanation for it was shown in Chapter 5 how the horizontal limestone 'scars' of the Yorkshire Dales are part of Northern England's most beautiful countryside. There, instead of dour dark slate beds, the white limestones glitter on the slopes, their rich limy soils being clothed with verdant grass and colourful rowan. Above all, the limestone-built farms and fieldwalls reflect the light and project their image brightly onto the hillsides, while the dour slated cottages and their dark stone fieldwalls recede into the sombre colouring of the shadowy slate mountain. The distinction may lie subconsciously behind an atavistic choice that one makes between, on the one hand, the traditional handiwork of the countryman working harmoniously within the natural landscape and, on the other, the environmental discordance produced when modern industrialists exploit the land's resources to such a degree that the surface scars cannot be healed. On a small scale slates have been quarried in Britain since Roman times and some of the earliest roughly hewn slates have helped to adorn many of our most historic buildings. It was only when the canals and railways of nineteenth-century Britain allowed slate to be transported cheaply to the burgeoning cities that mass production brought spoliation to the mountains and uniformity to the urban roofscapes. Nowhere was this transformation more apparent than in North Wales.

## Slate in North Wales

The slate mountains of Snowdonia comprise the majority of the hump-backed satellites which crowd round the soaring majesty of Snowdon itself. From its towering summit on a clear day it is possible to discern derelict slate workings to north, west and south; only the bulk of the intervening ridges of Y Garn away to the north-east hide the prodigious Penrhyn quarry at nearby Bethesda. The main slate belt of North Wales occurs in a narrow 15-mile tract of foothills which buttress the north-western shoulders of the highest peaks, and anyone entering the mountain fastness from either Bangor or Caernarfon cannot fail but to witness the way in which the slaty rocks have been exploited. And not only by the hand of man. More than 20,000 years ago all the Snowdonian valleys were choked with glaciers which slowly bulldozed their separate ways down to the surrounding plains. Spawned in the highest cwms of the central massif, the glaciers hacked and chiselled the volcanic rocks into the jagged and spectacular terrain that one sees today. But the volcanic lavas and other igneous rocks were tough enough to constrict the glaciers to such an extent that the upland valleys are relatively narrow despite their considerable depth. Once the moving ice reached the slate belt, however, it was a

Slate tips and
field barn
beneath
Snowdon. Note
the variety of
ways that slate
has been used.

different story, for these less resistant rocks allowed the glaciers to carve wider valleys just as they debouched on to the lowlands. This explains why the three main slate quarries of the north-western flanks all occur at the very portals of the mountainland, almost as if their stony faces were the monolithic slaty gateposts of the fortress itself.

From Bangor, the journey southwards along the A5, Telford's great Holyhead road, brings the traveller to the jaws of the mighty glacial trough of Nant Ffrancon, where stands Bethesda, the home of the enormous Penrhyn quarry, Britain's largest artificial void. The gentle slope of the mountain has been torn away in less than two hundred years to create an artificial working-face well over a mile in length and 1,200 feet in height. The twenty-one separate galleries have exposed the mountain's bones of grey, green and purplish-blue rock, each terrace with its apron of waste – the *tomennydd rwbl* (rubble mounds) of the Welsh quarryman. Little has changed since Michael Faraday visited the quarry in 1819 and described how 'Smooth perpendicular planes of slate . . . appeared above and below; in all directions chasms yawned and precipices frowned. Natural precipices do not convey the idea excited here, because they are in part rounded by the weather. But here every fracture, whether large or small, presented sharp angles'. Today the quarry, which once employed 3,000 men, is relatively quiet, its working life almost ended as the demand for slate has dwindled. Though 97% of its output went for roofing, there was once a need for tombstones, paving slabs, steps, window sills, lintels, benches, shelves, troughs, fences and writing slates. Before modern damp courses were invented slate was also used as damp proofing in Victorian homes. All such requirements are no more and purple-roofed, grey-stoned Bethesda, a product of its own exploitation, stands somewhat bemused, unsure whether it can discover a new role as a tourist centre.

Llanberis, notwithstanding its gigantic Dinorwig quarry, has always been invaded by visitors because of its location at the foot of Wales's highest mountain. It has the added attraction of two picturesque valley lakes, occupying the hollow scoured out in the yielding slate belt by glaciers formerly constrained by volcanic rocks in the narrow confines of the awesome Llanberis Pass. The town puts on a more cheerful face as it gazes across their rippled waters to the climbing slate galleries now converted into Britain's largest pumped storage scheme for the production of hydro-electricity. The stricken faces of grey and blue slate stand forlorn and vacant above the widespread skirts of man-made scree. For every ton of finished slate no less than 20 tons of waste were thrown away.

The way in which the slaty beds, known as the 'slate vein' by the quarrymen, outcrop steeply on the valley sides at Bethesda and Llanberis ensured that the quarries would be opened on the mountain slope (see Figure 10). In the third of

A slate fence in North Wales.

these northern slate settlements, however, the slate has been worked partly in open pits on the floor of the Nantlle valley and partly on the flat top of the neighbouring hill, reflecting the different character of the outcrop of the valuable 'vein'. The scattered Nantlle quarries, less concentrated than those of Bethesda and Llanberis, are the oldest in North Wales and probably supplied the floorstones used in the Roman forts of Segontium (Caernarfon) and Caer Llugwy (near Capel Curig). One Nantlle quarry was certainly visited by Edward I, though his great stone castles at Conwy and Caernarfon, slated in 1287 and 1317 respectively, may have been roofed with Bethesda slates. The ancient town of Conwy, which claims the first recorded use of slate roofing on an urban scale, appears also to have been using Bethesda slates in 1399, though some were brought down the Conwy river from the inland quarries at Penmachno, far to the south of the major slate belt.

The slaty rocks of southern Snowdonia are not only younger than those already described, they are also less steeply dipping where they reach the surface. Thus, the slates could be won by driving tunnels obliquely into the hillside (see Figure 15). The major slate town of these parts is Blaenau Ffestiniog, embowered among slaty hills but also divorced from the rugged grandeur of the highest mountains. Although some surface quarries exist, the blue-grey slates are largely mined from underground caverns which follow the 'vein' deep into the heart of the mountains.

A quarryman's
farm,
Snowdonia.

Because the Blaenau Ffestiniog quarries have retained a slender workforce and a modest level of production, it is still possible to recapture something of the thriving slate industry that once brought noise, smoke and dust to this beleaguered settlement which is nothing if not a slate town. One quickly discovers that this is the last bastion of slate-quarrying in Wales and begins to understand why some fifty years ago Professor A.H. Dodd believed that 'slate-quarrying is in a sense the most Welsh of all Welsh industries. In the quarries themselves English is a foreign tongue, and those shapeless monotonous villages which flank them, grey and drab, but clean . . . are the homes of a vigorous national culture'. But since these words were written many of the quarry villages have been virtually abandoned as the demand for slate has disappeared. One such settlement was Cwmorthin, in a high valley overlooking Blaenau Ffestiniog. It was described by Thomas Pennant in 1760, before the slate 'vein' was opened, as a hollow 'containing a pretty lake and two tenements which yield only grass'. Two centuries later it is a ghost village where the eyeless windows of the chapel and the workers' terraced cottages stare vacantly over engine houses, offices, reservoirs and dams, all crumbling and decayed. This feeling of dereliction has infiltrated the town of Blaenau Ffestiniog because of the mountainous tips of waste which frame the view from every corner. Rainwater seeps from the man-made screes and drips from the mossless slated roofs, it sluices across slate-grey pavements, trickles through slaty fences and surges down past slatestone walls before finally encountering the peripheral

An ancient cottage, near Beddgelert, Gwynedd, built from volcanic blocks but roofed with slate.

verdure of this once pastoral valley. To reach the scanty mountain pasture above the pockmarked slopes, dusty-fleeced sheep clamber up the grassless tips or simply scavenge through Blaenau's treeless terraced streets. It would be difficult to find a bleaker, starker or stonier townscape anywhere in Britain; the very paving stones seem to ooze despair. Yet Blaenau Ffestiniog has made a virtue of necessity by opening its slate mines to the public. Tourists can watch skilled craftsmen split the blue-grey slate into a variety of different sized roofing slabs, all named after titled women. The smallest slates are termed 'ladies' but they range upwards through 'countesses' and 'duchesses' to 'queens' which are 3 feet by 2 feet in size. Alternatively, visitors can travel deep into the mountain on slate tramways or take a longer trip on the famous Ffestiniog railway. Passengers rather than slates now make the spectacular journey far down the mountainside to the specially built coastal slate quays at Porthmadog.

Away from the quarry towns the remainder of the Welsh slate country has little to distinguish it from the surrounding landscapes except for its maze of drystone walls that reticulate the hillsides. This patchwork of tiny fields, each with its stone-built cottage, extends far up the misty mountain slopes, carved out of the common land by quarrymen during the widespread enclosure movements at the end of the eighteenth century. As employment in the quarry remained uncertain, before slate extraction experienced its nineteenth-century boom, the workmen were glad to find alternative sustenance from their windswept smallholdings.

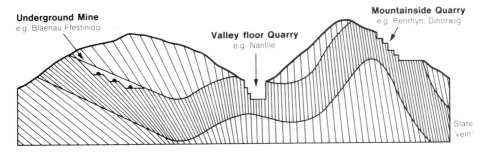

Today, this remarkable landscape of enclosure, incorporating a number of biblically-named hamlets such as Carmel and Nazareth, forms one of the most distinctive settlement pattern in North Wales. Though many of the drystone walls and the older cottages were built largely from glacially dumped boulders cleared from the fields, slabs of rudely fashioned slate wrenched from a local outcrop were also regularly used for rudimentary fences, doorposts, lintels and steps. In some parts of Snowdonia the use of slate as a building stone became more widespread in the nineteenth century, its wedge-shaped pieces, though needing plenty of mortar, could be bedded firmly together to give the sombre-coloured houses a weatherproof solidity in the river valleys of the Conwy, Llugwy, Lledr and Machno. But it is the medieval dwellings with their cyclopean mixture of volcanic and slaty masonry that remain in the memory. Such cottages as Ty Hyll (Ugly House) and Fedw Deg, both near Betws-y-Coed, are the most appealing, with the massive slate doorways of the latter inspiring a recent author to write that 'In its rugged magnificence and crudity it suggests irresistibly the entrance to a Megalithic tomb'.

## Slate landscapes in England and Scotland

Despite the fact that slate has long been quarried in the Lake District, Leicestershire, Cornwall, Devon and Scotland, the entire output from all their quarries was never more than one sixth of that from North Wales, so that their spoliation and dereliction is considerably less than in Snowdonia.

In the Lake District the slaty rocks, like those of North Wales, produce landforms of gentler contour and more rounded skyline than those created by their neighbouring volcanics. It is true that the Skiddaw slates build an eminence as high as the other Lake District giants but Skiddaw's slopes are smooth by comparison with those of Scafell. Despite the width of their outcrop the grey-brown Skiddaw slates have played only a minor role as a local roofing or building material. The main commercial slate belt of the Lake District lies to the south of its craggy mountainous core, extending in a narrow tract north-eastwards from the Duddon valley through the Tilberthwaite Fells, Elterwater and Ambleside to the Kirkstone Pass. In general this swathe of countryside, around the heads of Coniston Water and Windermere, is one of gently rolling hills, thick woodlands and picturesque rivers swirling through rocky, moss-smothered gorges. It is a landscape of considerable pastoral charm, possessing little of the bare austerity of the Snowdonian slate country, thanks mainly to the extensive broadleaf woodlands which have survived to grace the Cumbrian fells. Recent studies have shown that no less than 80% of Snowdonia's broadleaf woodlands are failing to regenerate, largely due to livestock grazing in a region where some two-thirds of the wooded area is open to stock. Because the Lakelander has always remained a farmer first and foremost and the slate industry has played a relatively minor role by comparison with that of North Wales, the Lake District quarries do not intrude upon the scene. Furthermore, with the exception of Elterwater village and its slatestone rubble buildings, Cumbria has nothing to compare with the Welsh slate towns or the mosaic of quarrymen's smallholdings which give the Snowdonian slate belt such an air of suspended animation, as if the frenzied activity of a century ago had been fossilized to match the tiny primeval creatures contained in the purple Welsh slates.

The true slates quarried in south-west Lakeland are blue-grey in colour while those taken from the small quarries to the west of Ambleside have a greenish-grey hue. These roofing slates, formerly known as Westmorland Slates, are thicker and heavier than Welsh slates, thereby necessitating considerably stouter beams to withstand their weight. Few spoil heaps mar the vistas within the Lake District valleys, largely because of the modest scale of production but partly because the slabby stone produced as a by-product of the roofing slate industry makes an excellent building stone for cottage and drystone wall alike. As in other Lakeland cottages, however, the majority of the slatestone dwellings have been liberally coated with whitewash, except in the towns and villages where the dark stonework brings a degree of dourness to the streets. The slatestone masonry is of two types, sharply angular or partly worn and sub-angular, depending on whether it has been picked from the quarry's waste tip or simply gathered from the nearby field or stream bed. Rarely does one find slatestone used in the form of large slabs in Cumbria's vernacular architecture; it is much more common to see only thinly cleaved stones roughly coursed and set in mortar which is kept well back from the wall surface. Thus most of the older Lakeland buildings are characterized by deep horizontal crevices between the slatestones where mosses and lichens have managed to gain a hold. The resulting textural roughness of the walls, therefore,

Honister Crag.
The last green
slate quarry in
Lakeland.

The Scotsman's House, Ambleside, Cumbria.

brings shadowy patches and niches which break up the monotonous greyness, though the large green slabs set endways above the doors and windows of Elterwater's quarrymen's cottages gives them a slightly bizarre and quizzical look. As far as bizarre slatestone houses are concerned, however, nothing can match the ingeniously designed 'Scotsman's House' built astride one of Ambleside's streams.

A Hartsop
cottage showing
the varied uses
of Lakeland
slate.

Elterwater slate quarry, Cumbria.

The slate quarried at Coniston, Tilberthwaite and Kirkstone Pass is arguably one of Britain's most attractively coloured rocks. Its subtle variations of olive green and sea green give an unforgettable hue to all the Lakeland buildings in which it has been used: 'When the sun breaks through after a shower of rain, these Cumberland Green roofs are a source of special delight' to the likes of Alec Clifton-Taylor. Today, the last green slate quarry, high up on the face of Honister Crag, produces only a limited supply of stone, to furnish thick cladding for modern city buildings or simply for domestic fireplaces; there is virtually no demand for its roofing slates, which is a matter of some regret. Paradoxically, this attractive stone is not a true slate but is a well cleaved, fine-grained volcanic ash laid down when the volcanoes were pouring out the lavas which now build Lakeland's most rugged mountains.

The most important source of slate in the English Midlands was the ancient quarry of Swithland in Leicestershire. Here in the Charnwood Forest, where a few modest hills of England's oldest rocks rise like islands above the plains and vales of

Slate cladding of houses is uncommon in Britain. This example is at Llechwedd, Snowdonia.

New Red Sandstone, the Romans found hard, well cleaved stones to roof and floor their villas and temples. When the Romans departed the quarries were to remain abandoned for a dozen centuries until their rediscovery in medieval times. Henceforth, their blue-grey and pinkish-purple slates were to be used for roofing throughout the East Midlands until production ceased in 1887. Nevertheless, the churches of Charnwood Forest and many churchyards in both Leicester and Loughborough retain some tombstones of Swithland Slate. The roofing slates, irregular in size and roughly hewn, were widely adopted to replace the less hardwearing thatch, though in turn the slates themselves are being replaced by modern tiles because their source of supply has disappeared; now the Swithland quarries stand forlorn, water filled and tree shrouded. Unless they are reopened it is possible that before long the only survival of these attractive building stones will ultimately be in the drystone walls which were constructed around 1812 when Charnwood was first enclosed.

England's only other major sources of true slate occur in the south-west, largely in Cornwall and Devon. Here, many of the rocks surrounding the granitic masses were transformed into slates by the heat and pressure associated with the formation of the granite. In general the slate landscape of the West Country is one of rolling hills, deeply incised valleys and flat coastal platforms. Slaty rocks have been quarried for centuries throughout Devon and Cornwall, usually in one-man quarries that were little more than surface scratchings. Only in a few places was this piecemeal exploitation organized into a commercial enterprise, with the largest and best known of these commercial quarries being that at Delabole in Cornwall which has been worked since the time of Elizabeth I. As early as 1602 Richard Carew was commenting that Cornish slate was '. . . in substance thinne, in colour faire, in weight light, in lasting strong', all qualities that made Delabole Slate greatly sought after as a roofing material. Its dark grey-green colouring has brought a certain distinction to many Cornish roofs, though near to the quarries slate stone has also been used for flooring, lintels, window sills and mantelshelves. Not surprisingly, the quarry village of Delabole is entirely slate-built, amidst more than 100 acres of spoil that have been tipped indiscriminately around England's largest slate quarry. Such spoliation led Betjeman to speak disparagingly of how '. . . the main road from Polzeath gives an arid view of this place which seems bleak with its grim church'. Yet, where slates have been utilized extensively in the West Country as a means of wall-cladding, few would dispute that they add an attractive element to the villages and townscapes. Slate-hanging, which is almost entirely restricted to Devon and Cornwall, was first used to protect the houses but later merely for decorative purposes. It became especially popular in Regency times and many slate-hung walls of this age have survived in such towns as Ashburton, Launceston, Padstow, Topsham and Totnes. One can still agree with Evans's 1812 description which noted that 'Houses thus fronted with a fine coloured slate, the slates well selected and put on neatly with black or dark grey mortar, assume a handsome appearance' (J. Evans). In contrast to the greenish tones of the Cornish slates, those from Devon are warm blue-browns and fawns which, despite their attractiveness, are not as durable as the Cornish stone. Nevertheless, all these West Country roofing slates are so roughly surfaced that they are able to retain substantial growths of moss and lichen, giving them a particularly appealing tone and texture which adds to their natural beauty. In this respect they are more comparable with the limestone roofing 'slates' of the Cotswolds than with the smooth, shiny, untextured slates from North Wales. Indeed, until the nineteenth century, a substantial proportion of the West Country roofing slate output was sent by sea to London and even to Windsor from such ports as Padstow and Fowey. Some of the smaller slate quarries were actually opened in the natural outcrops of the coastal cliffs and many of these abandoned workings can still be seen near Tintagel. But so severe are the gales on those exposed Cornish coastlands that many of the slate roofs have been slurried with cement or battened down with liberal applications of tar or red lead.

In Scotland slaty rocks are common in the Southern Uplands where they have been quarried sporadically near Peebles, but never on any substantial scale. The only commercial slate belt, however, occurs in a narrow tract along the southern shore of Loch Linnhe and the Firth of Lorne. Here, along the tattered coastline of western Scotland, the magnificent scenery is occasionally interrupted by the signs of quarrying. At the tiny village of Ballachulish, between Fort William and Oban, the grey-black slates have been hewn from a small coastal hill since 1697. They are harder and more resistant than most Scottish slates, which means that they can be cleaved thinner. At the height of production in the nineteenth century the quarry employed more than one quarter of the village's population but since the slate was shipped almost entirely by sea the remoteness and high freight charges meant that the quarry was destined for closure once clay and concrete roofing tiles were introduced.

Scotland's longest surviving quarries lay further south-west on the Slate Isles of Seil and Luing, two of the so-called Isles of the Sea. Here the slates are rough-textured and inset with gleaming yellow crystals of iron pyrites, known as 'fool's gold'. Because the specks of iron are the first to weather away, these slates are characterized initially by brown spots and later by tiny angular pits. Not surprisingly, because of their insular environment, the quarrymen were also lobster fishermen and farmers, diversifying their employment to meet the fluctuating demand for slate. Both Cullipool (Luing) and Easdale (Seil) have been built as quarry villages but, unlike their Welsh counterparts, their 'neat white cottages, roofed with dark island slate, stand in rows, with a jetty and dykes built of slates up-ended carrying on the geological formation of the rock' (*Shell Book of Cottages*). There is a record of Easdale slates being used at Falconer's Castle, Appin, as early as 1631 but, like the Ballachulish quarry, those of the Slate Isles were doomed, with the demise of Easdale being hastened when the sea broke into the shoreline quarry and drowned 240 men. Again, it was the isolation and high freight charges which, combined with the introduction of alternative roofing materials, were to signal the end of slate quarrying on a commercial scale in Scotland. The last major order from Easdale was to supply slates for the rebuilding of Iona Priory in the 1960s.

At the end of the nineteenth century four out of every five new houses built in

Britain were being roofed with slate. By 1938 clay or concrete tiles covered 80% of the newly built houses, partly due to cheapness but also because of fashion and a desire for more colourful materials. In addition, the tiles could be mass-produced rapidly by mechanical means whereas slate working, despite mechanization, remained a highly skilled and labour-intensive craft.

# PART III

# LANDMARKS OF STONE

# EIGHT

# STONE IN THE SERVICE OF MAN

For thousands of years man has used local pebbles, boulders, flagstones and rubble to construct the dwellings, fieldwalls, bridges and temples which give such character to the British landscape. Almost invariably the stone was utilized within a mile or so of where it was found on hillside, stream-bed, cliff-foot or shingle beach, largely because its bulkiness precluded long distance transport. Before the Norman Conquest stones were carried from distant sources only in exceptional circumstances: it has been seen how, for example, stone axes fashioned from Snowdonian and Lake District volcanic rocks were traded throughout Southern Britain and how a regular trade in flints once flourished along the ancient ridgeways of the English downlands. One of the most remarkable cases of stones being moved far from their point of origin concerns the famous 'blue' stones of Stonehenge. The majority of the mighty monoliths which build this unique landmark are sarsens dragged from the neighbouring downs, but scholars have long identified some of the innermost stones as having been brought from the Preseli Hills of North Pembrokeshire, a feat in which the remarkable ingenuity of our prehistoric forebears is worthy of comparison with that of the Egyptian pyramid-builders.

Nonetheless, such enterprises in primitive 'engineering' were exceptional and it is more normal to find only local materials employed in the rudimentary edifices which still adorn the countryside, from Cornwall's granite 'quoits' to Scotland's sandstone 'brochs'. Not until Roman times was Britain to experience the more sophisticated skills of quarryman and stonemason, though many of their landmarks have had to be excavated from beneath rural field and city street. The fine mosaics and accomplished masonry which have been unearthed at numerous Roman settlements suggest that substantial stone-built towns and villas once decorated the English scene. The Cotswold oolites provided the Romans with a particularly rich source of freestone, though local materials were incorporated in all the far-flung forts and road links of their military zone in upland Britain. Welsh slates, for example, were just as valuable to the masons of Roman Caernarfon as were the slabs of Millstone Grit to the legionary road-builders of the Pennines. To see Roman stonework in its grandest form, however, one must visit the remarkable bulwark of Hadrian's Wall.

Both the Vikings and the Anglo-Saxons contributed their share to the stone buildings of Britain but the surviving examples are rare since many remnants have been destroyed or incorporated into later structures. Furthermore, a great deal of Saxon architecture was based on wood rather than stone. The majority of their buildings were modest and, apart from the handful of Saxon stone churches, it has to be acknowledged that it was the Normans who revolutionized English stone-building. Their society initially required enormous cathedrals, churches and castles to proclaim its majestic dominance; later would follow the town house and market hall, the manor and farm, the cottage and barn, the windmill and viaduct. Because few skilled quarrymen and stonemasons were to be found among the indigenous population it became necessary to seek expertise from across the Channel and, not surprisingly, together with the craftsmen these early Normans also imported their tried and trusted building stone from Caen in Normandy. Many of the earliest Norman structures in southern England were constructed from Caen Stone (similar to our own Portland limestone) particularly where it could be supplied cheaply and easily by means of water transport.

It seems likely that, with the possible exception of Roman times, the Norman Conquest was the first occasion in the history of British building that rigorous principles were introduced into the choice of building stone. The questions that had to be answered included: what structural and architectural function was the stone expected to fulfil? What texture and colour was required? What were the local climatic conditions it would have to withstand? and last but by no means least: What were the costs that would be incurred? The structural requirements were as important to the quarryman as they were to the mason, for a building in which the stone had been carefully cut and laid as closely as possible to its naturally bedded condition was likely to last longer when exposed to the vagaries of the weather. There are many documented examples of stones that were badly quarried and wrongly laid, leading to early decay and ultimate collapse.

The properties that have to be judged when deciding upon the quality of a building stone can be grouped into three major headings: first, jointing and bedding, second, strength and durability, and third, texture and porosity. The spacing of the natural fissures in the bedrock, whether they be the bedding planes or the joint patterns, is of paramount importance, for closely spaced joints and thin beds may render the material useless as a building stone. Only in the case of flagstones for building purposes or where slates or stone tiles are required for roofing will a thinly bedded or highly cleaved rock be in demand. Most freestones have an easily worked joint pattern that helps to reduce the costs of quarrying because so little waste is produced. The strength and durability of a building stone depends not so much on the hardness of its individual grains as on the degree of bonding or aggregation of its particles. It follows that a well cemented sedimentary rock, such as a gritstone, will be tougher than a loosely cemented sandstone or limestone. But an igneous rock, such as a granite or dolerite, whose character has been tempered in the subterranean furnace and whose grains are virtually fused together, will be well nigh indestructible. Equally, its hardness will make it expensive to quarry and difficult to carve. Not surprisingly, therefore, igneous stone has been utilized for centuries in situations where functional durability is more important than appearance: bridges, lighthouses, embankments, harbours and road surfaces are the most obvious examples of such use, although where a durable, sturdy and steadfast appearance is required granite has long been favoured for certain banks and prisons. On turning to the texture and porosity of a potential building stone the quarryman knows that a rock with open pores allows water to sink in, thereby making it more liable to chemical decay and frost disintegration. He will also know that when a rock is first hewn its pores will retain 'quarry water' for several months before it dries out and that no reputable mason would build with unseasoned 'green' stone. After a spell of drying the exterior pores will have weathered sufficiently to create a type of crust or rind that not only helps to waterproof the stone but also serves to produce an aesthetically pleasing surface mellowness. The surface texture of a building stone is not simply a matter of its visual appeal, however, since the best quality freestones are characterized by a uniform grain size which leads not only to even rates of weathering but also to a greater facility of dressing and carving. Thus the even-grained limestones and sandstones have always been most popular for use in cathedrals, churches, palaces and stately homes, many of which were to be further adorned with richly carved statuary.

The Anglo-Saxons quarried their stone in three major areas; but Quarr in the Isle of Wight and Box in Wiltshire were never as important as Barnack in Northamptonshire. This famous quarry has been worked since the seventh century, its oolite being first shipped down the River Welland and up the East Anglian rivers to help build Norwich Cathedral and the Abbey at Bury St Edmunds. By Norman times the records abound with references to stone quarries and water transport, with building stone being exempted from tolls when it was freely given to religious communities. Because the Normans appear to have considered that Anglo-Saxon architecture was inferior the prelates quickly set about rebuilding the majority of the cathedrals. Those at Canterbury, Lincoln, St Albans, Rochester and Worcester were commenced before the Domesday Book was compiled in 1086, while Durham was the first major British building to be roofed with ribbed stone-vaulting. In addition to the cathedrals hundreds of new monasteries and perhaps thousands of parish churches were built or rebuilt before

1200, a use of stone never to be paralleled in England until the Victorian era. Moreover, William I, needing to impose his administrative and military authority upon a conquered nation, dazzled the peasantry with an array of stone-built castles to replace the existing structures of earth and timber. Surviving Roman buildings were pillaged for their worked stone and many Norman castles, such as Portchester and Wallingford, were built within existing Roman fortresses. To construct his major castle at Windsor, William had Totternhoe Stone transported from the chalk quarries of the nearby Chilterns, but for London's White Tower, Norman England's most important secular building, he preferred to import Caen Stone. Today, this early Norman masterpiece of glistening white stone contrasts with the few exposed remnants of the Roman city wall built from squared blocks of dark Kentish ragstone.

Throughout the Middle Ages the selection of stone for London's buildings was determined partly by the cost of transport and partly by the lack of efficient stone-working machinery. Consequently, only the churches were built of stone, the remainder of the buildings being raised from timber and brick and thatched with straw, all of which commodities could be supplied from local sources. After a series of disastrous conflagrations, however, Henry FitzAlwyn, Lord Mayor of London in Richard III's reign, decreed that no one should build within the city 'but of stone, with a roof of slate or burnt tile'. Though some English stone was shipped around the coast from Purbeck, considerable quantities of Caen Stone were still arriving from Normandy, especially to assist in the extensions of Westminster Abbey. Nonetheless, Chilmark Stone was brought from Wiltshire to build the Abbey's north transept and Chapter House, while another mayor, the famous Dick Whittington, contributed towards the cost of paving the Guildhall with Purbeck Marble. Yet the seventeenth century had dawned before English freestone was imported to Britain's capital city in anything other than sporadic amounts. It needed an architect of the stature of Inigo Jones to discover the qualities of Portland Stone and it was this gleaming white limestone that he first introduced to London's townscape in 1619 in the form of the graceful Banqueting Hall in Whitehall.

The small treeless Dorset peninsula known as Portland Island is composed of 3,000 acres of Britain's finest freestone. A modern visitor would be dismayed at the devastated landscape that has been created in less than four centuries as each of Portland's acres has yielded no less than 350,000 cubic feet of masonry. And since both the highly prized seven-foot thick oolitic layers (known respectively as the 'Whitbed' and the 'Basebed') are overlain by a less valuable shelly limestone (the 'Roach'), it is hardly surprising that 150 million tons of rubble waste now litter the scene. Though lacking the uniform texture of the oolites the Roach limestone has

The Banqueting Hall, Whitehall, London. The first major use of Portland Stone in London.

also been in demand, because its unyielding nature suited the harbour walls at Portsmouth Dockyard and Portland naval base and such quays as the famous Cobb at Lyme Regis. Nevertheless its shell casts and cavities have rendered it unsuitable for carving. Not so the dazzling oolitic limestone beds which lent themselves to intricate mouldings and engravings, for they possessed properties which satisfied all the design principles of the architect – grace, style, symmetry, colour and restraint. Yet they remained too hard to saw without expenditure of enormous time and labour which meant that not only had the Portland quarries remained undeveloped for centuries but they had failed to create a local landscape of stone buildings comparable with those of the Cotswolds or the Pennines. Most authorities agree that Portland Stone achieves its greatest impact in an urban context where, rather than blending with other building materials, it relies upon its snowy white colour to create a striking visual contrast.

The opportunity for this beautiful stone to exhibit its much vaunted attributes was not long in coming, for a few decades after Inigo Jones had first brought Portland Stone to London the centre of the capital was destroyed by the Great Fire of 1666. St Paul's Cathedral and 88 parish churches were lost in the conflagration, but thanks to the remarkable energy and architectural brilliance of Sir Christopher Wren a new St Paul's and 54 churches were constructed between 1675 and 1710. The chosen stone was almost invariably Portland limestone, brought by sea

The British
Museum,
London.

around the coast in vast quantities, until Wren's spires and steeples gleamed 'like lilies among the rose-red brick of Canaletto's paintings'.

In addition to the churches and the majestic edifice of St Paul's, the diameter of whose pillars was governed entirely by the size of the quarry blocks, Wren also constructed the well known limestone pillar of The Monument to commemorate the Great Fire. Though the western towers of Westminster Abbey were also built of Portland Stone in the early eighteenth century these were the work of Nicholas Hawksmoor, Wren's able pupil. The earliest of the Portland Stone buildings, which so dramatically changed the London skyline, were in the style of the Italian Renaissance, but demand for this dazzling white freestone was unabated as the capital city continued to grow throughout the nineteenth century. This was an era when architects favoured the Greek Revival style and such notable structures as the Bank of England, the Mansion House, the Custom House, the British Museum and the National Gallery added their stature to the London scene, all built from the graceful Portland limestone. Though granite was also making its presence felt in some of the City buildings a free offer of Scottish granite to rebuild the Houses of Parliament was turned down on the grounds of distance and difficulties of carving. The Commissioners chose instead a Magnesian Limestone from the East Midlands, an unfortunate choice (as will be shown below), although the Law Courts were sensibly built of Portland Stone in a Gothic Revival style much favoured by the late Victorians. The demand for Portland Stone continued throughout the twentieth century, with County Hall, Somerset House, the Reform Club, the Piccadilly Hotel, Bush House, much of London University and the Victoria and Albert Museum being among the best known examples of its immaculate character. Above all, the dazzling skyline of Whitehall and the Horse Guards, the epitome of London's corridors of power, owes much of its spectacle to the durable but elegant nature of the Portland limestone.

Since it was so successfully used by Georgian architects to build one of England's most beautiful cities, that of Bath, it appears somewhat puzzling that Britain's other great oolitic limestone, Bath Stone, was incorporated in London's buildings to a far lesser extent than that of its Dorset counterpart. Despite its warmer colour, Bath Stone is coarser and more open-textured than Portland Stone and its appears to have suffered rather more from the capital city's air pollution as a result. The main reason against it in the eighteenth century, however, was the fact that the Cotswold stone quarries were located far from any natural waterway save that of the Bristol Avon, when compared with the coastal sites of Purbeck and Portland. Thus, while Portland Stone could be transported relatively quickly and cheaply by sea, Bath Stone had to be taken laboriously and expensively overland until the building of the Kennet and Avon Canal between 1795 and 1810. Thereafter the creamy Cotswold oolite became increasingly adopted for London's buildings, particularly when it was chosen by the Duke of Wellington to construct Apsley House at Hyde Park Corner in 1828. In general it appears that while Portland Stone was reserved largely for churches, public buildings and commercial premises, Bath Stone went chiefly into private housing, particularly into the terraces of Regency London. Nonetheless, following the precedent set by St Bartholomew's Hospital (1730–59), which used Bath Stone long before it became fashionable in London, the Royal College of Physicians chose Bath rather than Portland Stone for its early-nineteenth century building (now Canada House in Trafalgar Square).

It was not only London that witnessed a rapid increase in the use of building stone during the heyday of Georgian fashion. Throughout the eighteenth and early nineteenth centuries there was an astonishing increase in the building or rebuilding of country houses throughout the length and breadth of Britain. In addition to the buildings themselves, their formal gardens, pavilions, statues, pagodas, temples, bridges and boundary walls were built almost invariably of stone, often transported at very great expense from distant quarries. Even the notable ruby brick façades of Georgian England had stone pediments, porches, window mouldings and other ornamentation. Nonetheless, until Victorian times local building materials were still widely used, giving to the towns and villages '. . . an inescapable aura of organic unity with their environment' (Reed). The building of the canals between 1760 and 1830 marked the beginning of the end of this vernacular tradition in building, while the railway age made stone not only more accessible but much easier and cheaper to transport. It must not be forgotten that the construction of the canals and railways also added further dimension to Britain's landscapes of stone, for the aqueducts, bridges and tunnels themselves provided Britain with some of her most monumental landmarks. Such features as the gritstone canal aqueduct at Lancaster (1797), Telford's memorable sandstone Pont Cysyllte near Llangollen (1805) and the sandstone Penistone railway viaduct in Yorkshire (1885) were thought to be spectacular achievements during the Industrial Revolution. But this wave of innovation also witnessed the invention of the Hoffmann Kiln in 1858 which enabled bricks to be mass-produced just at the time when Victorian cities were burgeoning and demand for cheap building materials was at its zenith. British cities were soon to become faceless and characterless gridirons of red brick and blue slate. Of the larger cities only London continued to exhibit any benefit from the improved transport facilities now available for building stone as was clearly illustrated by the increased demand for Bath Stone.

Once the Railway Age had connected the capital with the inland quarries of England's stone belt London's architects had a much greater freedom of choice for their building stones. Moreover, once Brunel had completed his Great Western Railway line to Bristol in 1841 he was able to take the valuable Bath Stone quarried from the Box railway tunnel near Bath on some of his first return rail journeys to the capital. Some of this material was probably incorporated in several of the late nineteenth-century London churches that were springing up in the rapidly developing suburbs, especially that of Kensington. Bath Stone was also used in the north and west fronts of Buckingham Palace, though its most famous front towards the Mall is of Portland Stone.

Although Bath Stone was not adopted as extensively as Portland Stone by London's architects, Kenneth Hudson believes that in the second half of the nineteenth century 'stone, especially Bath Stone, was the great respectability symbol, beloved by middle-class families for its fashionable echoes, by the builders for its cheapness, and by the banks, insurance companies, manufacturers and merchants as evidence of prosperity and commercial soundness.' The railways had finally given the Cotswold quarries the double advantage of more rapid transport and cheaper costs than could be enjoyed by the coastal Portland quarries. Thus, between 1850 and 1870 many urban architects preferred to use Bath Stone to enhance their townscapes. When it was finally decided to replace the crumbling edifice of Oxford's dreaming spires, for example, the University chose Bath Stone in place of its disastrous Headington Stone. This happened in the case of the well known stone heads that front the Sheldonian Theatre. A century later, however, the Bath Stone itself had succumbed to air pollution and, like the Houses of Parliament, has had to be restored with Clipsham Stone from the East Midland stone belt. By the 1930s the Bath Stone company had not only taken over the Portland quarries but had also acquired the other great limestone freestone quarries at Ham Hill, Doulting and Clipsham (see Figure 9) to create a virtual monopoly.

The only other limestone to be used on any great scale in London's architecture was the ill-fated Magnesian Limestone of the East Midlands. When the Government Commissioners decided to rebuild the Houses of Parliament in 1839, following a fire, they chose a limestone from Bolsover Moor in Derbyshire, though similar stone was finally taken from the Anston quarry near Sheffield. Unfortunately, the stone was poorly quarried and sent to London within a fortnight, allowing no time for its essential 'seasoning'. It was later noted by a Select Committee that *all* stone was taken, good, bad and indifferent, and that virtually no waste was left in the quarry, a sure sign that inferior stone was not discarded. The notorious air pollution of Victorian London rapidly turned the Magnesian Limestone into a mixture of Epsom Salts and gypsum once the sulphuric acid from countless coal fires had got to work on the stone. The commissioners, having seen the magnificent ornamentation of Southwell Minster in Nottinghamshire, carved in Magnesian Limestone from Mansfield Woodhouse, thought that this stone would also fulfil the requirements of the Palace of Westminster. A few stones from the same quarry were in fact used at Westminster and they have worn well, but by 1927 all the Anston Stone had been replaced by Stancliffe Stone, a sandstone from the Millstone Grit near Matlock.

Sandstone has had far less impact as a building stone in London than have the limestones described above. Though Kentish Ragstone from the Weald had been used in some of the city's earliest structures, there are few notable sandstone buildings still to be seen. Old London Bridge, built in 1176 from Reigate Stone (from the Upper Greensand), has long since disappeared, so too have virtually all the medieval churches built of Kentish Rag, destroyed mainly by the Great Fire, with the most eminent survivor being that of St Ethelburga in Bishopsgate. Millstone Grit, mainly from Darley Dale, was used in King's College Hospital and in the splendid Euston Arch (now scandalously demolished). Otherwise this tenacious Pennine stone was used more modestly to form the base of the Albert

Gritstone 'setts' in Main Street, Haworth, West Yorkshire.

Memorial and for part of the Thames Embankment. The warmer-toned sandstones from the Old Red and New Red are uncommon, the former being best exemplified in part of Brompton Oratory while the latter (from Dumfries) helps build Liverpool Street Station and part of Burlington House. Blocks of Aislaby Sandstone, brought by sea from Whitby, once formed the foundations of the Old Waterloo Bridge, now replaced by a glittering span of Portland Stone. Finally, the most widely used sandstone in London, as in many other cities, was the so-called York Stone which, because it could be split into flagstones, was adopted extensively by the Victorians to create their pavements. Because sandstone is more gritty than limestone it provides a more durable, non-slip surface for outdoor use. Though Purbeck limestone was popular for internal paving it was the hard York Stone, shipped first by canal and later by rail from quarries near Leeds, Halifax and Huddersfield, that formed most urban pavements throughout the Victorian era. This thinly laminated carboniferous sandstone has the added property that its 'peeling' layers create a slightly uneven surface which greatly assists pedestrians during wet weather. Similar stone has paved other British cities, from quarries in South Wales, the Forest of Dean, Macclesfield, Wigan,

Derbyshire and Northumberland. Furthermore Caithness flags, from the Old Red Sandstone, were occasionally shipped from the far north to pave the industrial cities of Lowland Scotland.

So far as the urban road surfaces were concerned cobbles were first used, some dating back to the Middle Ages. In the chalkland towns and villages flint was once ubiquitous, as illustrated by such fine old streets as Elm Hill in Norwich, while in the sandstone tracts of Midland England, the almost indestructible Bunter pebbles from the New Red Sandstone can still be seen in a few ancient streets and alleyways. One of the most striking examples of Bunter pebbles being used to surface an urban square, however, is that surrounding the Radcliffe Camera in Oxford. Here the pink carpet of uneven cobbles serves only to emphasize the smooth ashlar of the towering limestone buildings. Once the Industrial Revolution had brought mills and factories to the urban scene and had filled the city streets with wheeled transport a more efficient means of road surfacing was needed to withstand the commercial wear and tear. What better than granite to fulfil this requirement, for by the nineteenth century machinery had been introduced which could trim the obdurate stone with a speed and cheapness essential to meet the enormous demand. Before long many of the unsurfaced urban roads were systematically sealed with so-called granite 'setts' of variously sized, roughly dressed rectangular blocks, the dimensions of which were usually less than $7'' \times 4'' \times 5''$. Aberdeen was the principal supplier of the London market but granite setts were also brought from Mountsorrel (Leicestershire), Cornwall or Lundy Island. In some cases other igneous rocks, such as dolerite or basalt, were utilized but granite was the most widely adopted. Setts provided an excellent footing for horse traffic and, though many major cities have by now replaced setts entirely with tarmacadam, several examples survive in some London mews near to the former stable blocks. Since the noise and vibration of steel-shod wheels and hooves on granite blocks was almost intolerable, several London thoroughfares went through a noise-reducing phase when wooden blocks coated with tar were introduced before pneumatic tyres and asphalt led to the demise of setts altogether.

Granite has survived in the townscape in many other ways, not least in Britain's capital city where it was first used by Rennie in the original Waterloo Bridge as early as 1817. Though this particular granite came from Aberdeen's Rubislaw quarry, the new London Bridge (1831) was built from Dartmoor granite and the foundations of the Houses of Parliament (1840) from a similar granite hewn from the Hay Tor quarries. Dartmoor granite became especially popular during the latter half of the nineteenth century not only to complete the Thames Embankment (1865–85) but also to aid in the construction of the bridges

*(Above)* Sandstone Saxon crosses and Bunter pebble cobbles, Sandbach, Cheshire.

*(Above right)* Keele University, Staffordshire. A courtyard floored with gritstone *(foreground)* and granite setts *(background)*.

at Battersea, Putney, Kew, Vauxhall, Blackfriars and in Tower Bridge itself. The Foggintor quarry on Dartmoor was singled out to supply the prestigious blocks from which Nelson's Column was raised in Trafalgar Square. It was not only the increased skill of the mason and availability of more sophisticated stone-cutting machinery but also the rapid spread of the British railway network that finally gave West Country and Scottish granite a realistic opportunity to compete as a building stone in Victorian England. Furthermore, its highly polished crystalline character appealed to the contemporary architects and builders who were attracted for example, by the large white crystals exhibited by the Lamorna Cove granite, by the silvery-grey of the Aberdeen stone, the salmon-pink tones of the Peterhead granite or by the large flesh-coloured crystalline rock from Shap. Rather appropriately the silver granite from Scotland's Royal Deeside was used in the base of Queen Victoria's monument opposite Buckingham Palace and also in Trafalgar Square. But for the foundations of Buckingham Palace itself and for the lower courses of County Hall the granite was brought from Cornwall. The City saw fit to found the Royal Exchange and the Royal Bank of Scotland on a firm base of Bodmin Granite from the De Lank quarry, and not to be outdone the brokers from the New Stock Exchange in Threadneedle Street chose Bodmin Granite from Hantergantick quarry to give their structure a robust start. One of the most elegant buildings of the City, however, is Thomas Gray House (Number 39 Cornhill) where De Lank, Rubislaw and Peterhead granites have been skilfully combined. Most of the reddish-pink granite incorporated in London's buildings, such as that at the Carlton Club or in the base of the Prudential offices in Holborn, has been quarried at Peterhead, though the polished stone posts that ring St Paul's noble edifice are examples of Shap Granite. The polished blacks, garish pinks and other startling hues found in modern London's stone façades are largely imported from overseas and need no further comment here. The *Daily Mirror* offices have been clad in some fine Cumbrian green slate but, in general, the use of stone in London's post-war townscape has virtually ceased, especially once the ravages of the Blitz had been made good. Of Wren's magnificent churches only twenty-three have survived intact, though a further six towers are still extant. Today, glass, concrete and steel reign supreme and skyscrapers have relegated some of Britain's finest urban stonework to sunless canyons: London's famous skyline is now dwarfed by the newly-found image of 'High Tech' Brutalism. A few modern London buildings have been faced with stone but on the whole they make little impact on the urban scene in which artificial materials are paramount.

Another eminent British city with a townscape predominantly of stone is that of Edinburgh, one of the world's most elegant cities and certainly one of the 'stoniest'. Dark rocky cliffs loom above every street and punctuate every vista; the miscellany of hills and hollows is linked by a mesh of cobbled lanes, stone bridges and steps into a well integrated but complex urban organism. Above all, one senses that the unified sense of place that the city engenders is due in no small measure to the presence of the ubiquitous freestone fabric in which 'Auld Reekie' has clothed itself, a fine-grained sandstone which, by comparison with London's famous limestones, has survived the rigours of air pollution virtually unscathed.

Edinburgh's ancient citadel was sited on a lofty crag of dark volcanic rock similar to those neighbouring craggy eminences which represent the denuded remnants of a once mighty volcano: Arthur's Seat is its rugged but eroded stump; the jutting face of Salisbury Crags was formed from a sill every bit as striking as the Great Whin Sill described in Chapter 4; the lesser bulk of Calton Hill is simply an isolated lava knob. The modern townscape now laps round their stony flanks and utilizes some of their prominent summits to display its collection of sandstone monuments.

Unlike London, Scotland's premier city had good building stone close at hand, with the earliest builders making considerable use of the dark volcanic stone to raise the initial fortifications on the Castle Rock. But because the volcanic 'whin' was too hard to dress successfully it was soon relegated to rudimentary rubble walling, most of which has subsequently disappeared beneath a rendering of lime-based stucco known as 'harling'. So far as other building stones are concerned Scotland has plenty of granite which has been used extensively as setts to pave its city streets, but there are few useful limestones and certainly nothing comparable with the elegant oolites of southern England. Moreover, because it is so pebbly, the Scottish Millstone Grit has never figured so successfully as a building stone as it has in the Pennines. Nonetheless, the Scottish lowlands have a wealth of Old Red, New Red and Carboniferous sandstones which have provided the urban masons with almost inexhaustable supplies of excellent building stone. Dumfries and Ayrshire have many fine pink sandstone buildings of stone taken from their New Red Sandstone quarries, while Tayside boasts an equally fine crop of Old Red structures, including the ancient but well-preserved defensive round tower at Newburgh. The most famous of all the freestones, however, is the Carboniferous sandstone known as the Craigleith Stone. It underlies most of Edinburgh's classical 'New Town' and was quarried a few miles west of the city centre. It has been claimed that Craigleith Stone is to Edinburgh what Portland Stone is to London, Pentelic Marble to Athens and Pietra Serena to Florence. Generations of writers have eulogized about the qualities of Edinburgh's unique townscape, no doubt reflecting that the choice of its fine-textured, buff-coloured sandstone has given to the city an air of congruity rarely found in British towns away from the Cotswold stonebelt. It has to be remembered, however, that the city was also fortunate in possessing architects and planners who built with vision, grace and good sense.

Even before its mid-eighteenth-century expansion Edinburgh had been favoured with some architectural gems of Scottish Craigleith stonework, including St Giles cathedral (1450–1500), Greyfriars Church (1620) and Heriot's Hospital (1628–59), though its finest townscape remained the serrated skyline of the 'Old Town' which strides along the crest of the mile-long ridge that links the dour dark Castle with the cream-coloured classical façade of the Palace of Holyroodhouse. The Old Town's linear plan has been likened to a spine from which the ancient streets and alleys (the 'wynds' and 'pends') protrude like ribs down the ridge's steep flanks. The glaciers which steepened these slopes gouged out the softer rocks into linear hollows, that to the south now being infilled with streets whose names are redolent of the Middle Ages – the Grassmarket and Cowgate being two of the best known. Until the eighteenth century the northern trough was occupied by the North Loch but this has now been drained and

West Bow, Edinburgh. A 17th century 'street canyon', (from an 1839 engraving by T. Allom). Compare this picture with plate on p. 85.

Calton Hill, Edinburgh, showing the buildings and monuments of Craigleith sandstone.

replaced by the Waverley Gardens. Such reclamation reflected the need for urban expansion as the population continued to increase on the desperately overcrowded ridge of the Old Town. So congested was the site that the High Street's eight-storey tenements became Britain's first street canyons. The need for the city to spread northwards became paramount, not only to provide relief from the crumbling, unsanitary tenements, but also to make up for the dearth of public buildings. The outcome was the classical 'New Town', planned by James Craig in 1652 and executed by such eminent architects as the Adam brothers and John Playfair.

Today, Edinburgh's Georgian terraces and squares rank alongside those of Bath as the finest examples of classical urban form in Britain. Architects have ascribed its combination of order and spaciousness not only to the good proportions of the streets and buildings but also to the quality of the Craigleith Stone. Though a few structures, such as the Royal Scottish Academy (1828), were raised from stone shipped across the Forth from Fife, the majority of the public buildings, including the Register House (1780), the Royal High School (1829) and the University (started in 1789), were built from Craigleith Stone. Some critics have claimed that the latter building was Robert Adam's noblest achievement and one marvels at the audacity of its eastern portico supported on six monolithic columns, each carved from a single block of stone 22 feet in height and 3 feet in diameter! In conclusion, the Parthenon-style buildings on Calton

Hill explain the city's soubriquet as 'The Athens of the North' and also epitomize the magnificent decorative stonework of Georgian Edinburgh where masonry blends with native rock. Between 1776 and 1830 the bare summit of Calton Hill became, in the words of A.J. Youngson, 'littered with buildings of one sort or another, mostly of classical shape or form, a seemingly fanciful arrangement of harmonious composition which forms a pattern constantly and wonderfully changing as one looks up or across at the hill from different parts of the city.'

# SOURCES MENTIONED IN THE TEXT

*Page*

41. Betjeman, John, *Cornwall. A Shell Guide* (Faber, London, 1964)

43. Carew, Richard, *Survey of Cornwall* (1602)

48. Hoskins, W.G., *Devon: A new Survey of England* (Collins, London, 1954)

48. Penoyre, John & Jane, *Houses in the Landscape: A Regional Study of Vernacular Building Styles in England & Wales* (Faber, London, 1978)

50. Poucher, W.A., *A Camera in the Cairngorms* (Chapman & Hall, London, 1947)

57. Coleridge, S.T., *Tour in the Lake Country* (1802)

57. Defoe, Daniel, *A Tour through the whole island of Great Britain* (1724–6; republished Penguin Books, Harmondsworth, 1971)

61. Wordsworth, William, *Guide to the Lakes* (1835; reprinted with an introduction by E. De Sélincourt, O.U.P., Oxford, 1977)

66. Darling, F.F., *West Highland Survey: An Essay in Human Ecology* (Oxford University Press, Oxford, 1955)

72. Chapman, R.W. (ed.), *Johnson and Boswell: A Journey to the Western Islands of Scotland: The Journal of a Tour to the Hebrides* (Oxford University Press, Oxford, 1970)

72. Wordsworth, Dorothy, *Recollections of a Tour made in Scotland, 1803* (1805)

78. Rhodes, E., *Peak Scenery, or the Derbyshire Tourist* (1878)

81. Walton, Izaak, *The Compleat Angler* (1653)

84. Gray, Thomas, *Journal in the Lakes* (1775)

88. Young, Andrew, *A Prospect of Britain* (Harper, New York, no date)

89. Priestley, J.B., *English Journey* (Heinemann, London, 1934)

90. Hardy, Thomas, *The Well-Beloved* (1897)

91, 96. Hawkes, Jacquetta, *A Land* (Cresset Press, 1951; reprinted Pelican Books, Harmondsworth, 1959)

92. Evans, H.A., No title (1905)

96, 127. Wright, G., *The Stone villages of Britain* (David & Charles, Newton Abbot, 1985)

106. Thomas, Edward, *The South Country* (No date)

112. Leland, J., *Itinerary* (1710–12)

112. Defoe, Daniel, *op. cit.*

149. Dodd, A.H., *The Industrial Revolution in North Wales* (University of Wales Press, Cardiff, 1933)

149. Pennant, T., *A Tour in Wales* (1784; London, 3rd edn, 1810)

151. Scott, R., in North, F.J., Campbell, B. and Scott, R., *Snowdonia: The National Park of North Wales* (Collins, London, 1949)

158. Carew, R., *op. cit.*

158. Betjemen, J. *op. cit.*

158. Evans, J., No title (1812)

159. Reid, R., *The Shell Book of Cottages* (David & Charles, Newton Abbot, 1977)

170. Reed, M., *The Georgian Triumph, 1700–1830* (Routledge & Kegan Paul, London, 1983)

170. Hudson, K., *The Fashionable Stone* (Adams & Dart, Bath, 1971)

174. Youngson, A.J., *The Making of Classical Edinburgh* (1966)

OTHER BOOKS OF INTEREST

Anderson, J.G.C., *The Granites of Scotland* (Edinburgh, 1939)

Arkell, W.J., *Oxford Stone* (Faber, London, 1947)

Brunskill, R.W., *Illustrated Handbook of Vernacular Architecture* (Faber, London, 1970)

Clifton-Taylor, A., *The Pattern of English Building* (Faber, London, 1962)

Lindsay, J., *A History of the North Wales Slate Industry* (David & Charles, Newton Abbot, 1974)

North, F.J., *The Slates of North Wales* (National Museum of Wales, Cardiff, 1946)

O'Neill, H., *Stone for Building* (Heinemann, London, 1965)

Robinson, E., *London: Illustrated Geological Walks; 1. The City; 2. The West End* (Geologists' Association, 1985)

Shepherd, W., *Flint: Its Origins, Properties and Uses* (Faber, London, 1972)

Shore, B.C.G., *The Stones of Britain* (Leonard Hill, 1957)

Trueman, A.E., *Geology and Scenery in England and Wales* (Revised edn. by Whittow, J.B., and Hardy, J.R.; Penguin Books, Harmondsworth, 1971)

Whittow, J.B., *Geology and Scenery in Scotland* (Penguin Books, Harmondsworth, 1977)

# INDEX